Managing and Marketing for Pasture-Based Livestock Production

Edited by:

Edward B. Rayburn

Written by:

Geoffrey A. Benson	Thomas R. McConnell
E. Ann Clark	Bill Murphy
Julian Drelich, Jr.	Wesley N. Musser
Darrell L. Emmick	Phillip I. Osborne
Stephen A. Ford	James Y. Pritchard
David Grusenmeyer	Edward B. Rayburn
Marvin H. Hall	R. Mark Sulc
James S. Hill	Lester R. Vough
Lisa Holden	

Natural Resource, Agriculture, and Engineering Service (NRAES)
Cooperative Extension
PO Box 4557
Ithaca, New York 14852-4557

NRAES–174

January 2006

© 2006 by NRAES — Natural Resource, Agriculture, and Engineering Service

All rights reserved. Inquiries invited. (607) 255-7654

ISBN-13: 978-0-935817-99-7

ISBN-10: 0-935817-99-9

Library of Congress Cataloging-in-Publication Data

Managing and marketing for pasture-based livestock production /
    edited by Edward B. Rayburn ; written by Geoffrey A. Benson ... [et al.].
      p. cm. — (NRAES ; 174)
    Includes bibliographical references.
    ISBN 0-935817-99-9 (pbk.)
      1. Livestock. 2. Pastoral systems. 3. Animal industry. I. Rayburn, Edward B.
      II. Benson, G. A. (Geoffrey Alan). III. NRAES (Series) ; 174.

SF61.M24 2005
636—dc22   2005043867

Natural Resource, Agriculture, and Engineering Service (NRAES)
Cooperative Extension • PO Box 4557
Ithaca, New York 14852-4557
Phone: (607) 255-7654 • Fax: (607) 254-8770
E-mail: NRAES@CORNELL.EDU • Web site: WWW.NRAES.ORG

Table of Contents

ABOUT THE AUTHORS

CHAPTER 1: AN INTRODUCTION TO PASTURE-BASED LIVESTOCK PRODUCTION

Edward B. Rayburn
    Extension Forage Agronomist
    West Virginia University

Bill Murphy
    Department of Plant and Soil Science, retired
    University of Vermont

E. Ann Clark
    Associate Professor
    Department of Plant Agriculture
    University of Guelph

CHAPTER 2: VISION, MISSION, AND GOALS

Lisa Holden
    Associate Professor
    Department of Dairy and
    Animal Science
    The Pennsylvania State University

David Grusenmeyer
    Senior Extension Associate
    PRO-DAIRY
    Cornell University

CHAPTER 3: RESOURCE INVENTORIES IN FARM PLANNING

Darrell L. Emmick
    State Grazing Land Management Specialist
    U.S. Department of Agriculture,
    Natural Resources Conservation Service,
    New York

Julian Drelich, Jr.
    Resource Conservationist
    U.S. Department of Agriculture,
    Natural Resources Conservation Service,
    New York

James S. Hill
    Grassland Specialist
    U.S. Department of Agriculture,
    Natural Resources Conservation Service,
    West Virginia, retired

CHAPTER 4: ALLOCATION OF FARM RESOURCES

Stephen A. Ford
    Manager
    Blythe Cotton Company

Wesley N. Musser
    Professor and Farm Management Specialist
    University of Maryland

CHAPTER 5: MARKETING COMMERCIAL FEEDER CATTLE

Phillip I. Osborne
    Extension Specialist
    Animal Husbandry
    West Virginia University

James Y. Pritchard
    Pocahontas County Extension Agent
    West Virginia University

CHAPTER 6: DAIRY MARKETING

Geoffrey A. Benson
    Extension Economist
    Department of Agricultural and
    Resource Economics
    North Carolina State University

CHAPTER 7: DIRECT MARKETING

Thomas R. McConnell
    Extension Specialist
    Farm Management
    West Virginia University

CHAPTER 8: HAY MARKETING

R. Mark Sulc
    Associate Professor
    Department of Horticulture
    and Crop Science
    The Ohio State University

Marvin H. Hall
    Professor of Forage Management
    Department of Crop & Soil Sciences
    The Pennsylvania State University

Lester R. Vough
    Associate Professor and
    Extension Specialist
    Forage Systems Management
    University of Maryland

DISCLAIMER

The authors, editor, and NRAES cannot guarantee or warrant recommendations or products mentioned in this book. Use of a product name does not imply endorsement of the product to the exclusion of others that may be suitable.

Web site addresses listed in this book are current as of December 2005, but due to the dynamic nature of the Internet, some addresses may no longer be active.

ACKNOWLEDGMENTS

The authors wish to thank the following peer reviewers for offering comments to improve the quality and accuracy of the text. Note that some of the peer reviewers are producers.

Michael J. Baker
  Beef Cattle Extension Specialist
  Cornell University

Donald M. Ball
  Extension Agronomist
  Auburn University

John W. Berry
  Agricultural Marketing Educator
  Penn State Cooperative Extension,
  Lehigh County

Bill Bivens
  Extension Agent
  Michigan State University Extension,
  retired

Stephen L. Boyles
  Beef Extension Specialist
  Ohio State University Extension

Gary C. Burley
  East Hill Farms

E. Ann Clark
  Associate Professor
  Department of Plant Agriculture
  University of Guelph

Sister Augusta Collins, O.S.B.
  Agronomist
  The Abbey of Regina Laudis

Robert A. Cropp
  Professor Emeritus, Dairy Marketing
  University of Wisconsin–Madison

Stephen Derrenbacher
  Veterinarian
  Ruth Ann's Garden Style Beef

Sam Dixon
  Dairy Manager
  Shelburne Farms

John H. Fike
  Assistant Professor
  Forage-Livestock Research
  Virginia Polytechnic Institute
  and State University

David L. Greene
  Principal Agent Emeritus
  University of Maryland, College of
  Agriculture and Natural Resources

Hal Harris
  Professor
  Department of Agricultural and Applied
  Economics, Clemson University

D. W. Hartman
  Extension Agent
  Penn State Cooperative Extension

Duncan L. Hilchey
  Senior Extension Associate
  Community Food and Agriculture
  Program, Cornell Cooperative Extension

James S. Hill
  Grassland Specialist
  U.S. Department of Agriculture,
  Natural Resources Conservation Service,
  West Virginia, retired

Ronald J. Hoover
  On-Farm Research Coordinator
  The Pennsylvania State University

Gary W. Hornbaker
  Extension Agent
  Virginia Cooperative Extension

Jeffrey Hyde
  Assistant Professor of
  Agricultural Economics
  The Pennsylvania State University

John W. Irwin
  Extension Animal Scientist
  Clemson University Cooperative
  Extension

Glenn D. Johnson
  U.S. Department of Agriculture,
  Natural Resources Conservation Service/
  Virginia Polytechnic Institute and
  State University

Richard Kersbergen
  Extension Educator
  University of Maine Cooperative
  Extension

Wayne Knoblauch
  Professor
  Department of Applied Economics
  and Management, Cornell University

Ed Koncle
  Farm Seed Sales Manager
  Rohrer Seeds

Richard Leep
  Professor and Extension Specialist, Forages
  Department of Crop and Soil Sciences
  Michigan State University

Larry Lohr
  Dairy Farmer
  Cold Ridge Farms

Dale Miller
  Lecturer, Extension Specialist
  Animal Science Department
  North Carolina State University

Rory Miller
    Sheep Producer
Robert R. Peters
    Professor and Extension Dairy Specialist
    University of Maryland
Robert Schwart
    Professor and Extension Specialist
    in Dairy Marketing
    Department of Agricultural Economics
    Texas Cooperative Extension
Jane Shaw
    Soil Conservationist
    U.S. Department of Agriculture,
    Natural Resources Conservation Service,
    Virginia
Mark Stephenson
    Senior Extension Associate
    Cornell Program on Dairy Markets
    and Policy
Richard E. Stup
    Senior Extension Associate
    Penn State Dairy Alliance
Richard Swartzentruber
    Farmer Representative for Delaware

Peter Tozer
    Research Economist
    Western Australian Department
    of Agriculture
Dan Undersander
    Forage Agronomist
    University of Wisconsin–Madison
Dick Warner
    Producer and Chair, Grazing Land
    Conservation Initiative
Donald R. Wild
    Conservation Agronomist/Area Grazing
    Specialist
    U.S. Department of Agriculture,
    Natural Resources Conservation Service,
    New York

The authors thank Marty Sailus, NRAES director, and Cathleen Walker, NRAES production manager, for supporting this project; Joy Drohan, of Eco-Write, LLC, for copyediting and project management; Denise Plouffe, of Griffin Publishing, for layout and design; and Jeff Miller, of The Art Department, for preparing some of the illustrations.

Chapter 1
An Introduction to Pasture-Based Livestock Production

Edward B. Rayburn, Bill Murphy, and E. Ann Clark

Pastureland is an enormous resource that covers 6.2 million acres in the Northeast; when combined with rangeland, it covers 658 million acres, or 29% of the total area of the United States (25). Each ruminant livestock operation is based on a forage-livestock system, which encompasses forage species; animal species; and classes, management, and markets used by the producer to meet established goals. Some of the components of a forage-livestock system come with the operation; the manager may pick and choose others from a smorgasbord of alternatives.

Forage-livestock systems are multidimensional and need to be viewed from different perspectives. The three dimensions we will look at are:

1. the framework of management decisions around which a forage-livestock production system is developed (vision, goals/plans, resources, and markets);

2. the production centers through which the ecological flow of solar energy and cycles of water and minerals occur; and

3. the sustainability of the systems.

These three dimensions are of equal importance. Keep them in mind as you build your knowledge of pasture-based livestock production and implement concepts on the farm.

FRAMEWORK OF FORAGE-LIVESTOCK PRODUCTION

The framework of a farming system is manager- and site-specific. It is based on the manager's vision and goals and the plans made for achieving those goals using the available resources and markets. The ability of the manager to make decisions and work in an open market is the basis of the free-enterprise system. Production and marketing decisions are made at the management level.

Vision is a foreseen destination. Vision is knowing where you want yourself, your family, and your farm to be five to ten years from now. We all need a vision of what we want from life. Preferably, we should write it down and review it on a regular basis; otherwise, "If you don't know where you are going, any road will get you there."

Visioning is an important way to see how the farm operation fits into the rest of life, because it answers the question, "How do my activities and commitments in the home and community (school, church, local organizations, etc.) relate to my vision of life and my vision of the farm?"

Visioning must include all members of a management team, as teamwork is best when all team members share in developing the vision. All team members have a vision of what they want or need, so the vision must help all individuals achieve their own vision. Personal values play a big part in determining your vision, and your vision will change with your experience, accomplishments, and stage in life.

Goals are mileposts that must be reached to realize the vision. Goals are more specific than the vision. They focus management action by specifying what, when, and where things will be done. Goals define how to get to where you want to be, and they should be viewed in terms of personal, family, and business plans. You should have measurable long-term and short-term goals. Goals are part of the process of "beginning with the end in mind" (6).

Plans are specific actions (roads) taken to reach the goals (mileposts). Goals and plans together constitute a map leading to the realization of the vision. Planning occurs in a recurring cycle and includes how things will be done and who will do them. Like goals, plans can be divided into long-term (strategic) and short-term (tactical) operating plans.

Vision, goals, and plans are needed to focus management efforts. With focus, working out the plan will be more successful. One of Solomon's proverbs states that "Where there is no vision, the people perish…" (Proverbs 29:18). If the management team does not have a well-defined vision, then goal setting, planning, and daily work are not likely to be focused.

Resources are the things you have available to work with to accomplish your plans. Resources include management ability, capital (land, buildings, livestock, and equipment), and labor. A resource inventory allows the manager to view the resources available and evaluate the strengths and weaknesses of the resource base, the opportunities available to the operation, and possible threats from outside factors.

Markets provide the economic fuel that drives the plan. Dictionary definitions of market and marketing relate to the location and action of selling items. In the modern context, a major part of marketing is the planned production of an item having quality characteristics desired by an identified customer. It also includes developing new products that meet the present or future needs of customers and educational efforts to show real needs or to develop perceived needs for a specific product.

What the market wants should determine livestock type and management. Markets can be divided into established commodity markets and potential or developed niche markets.

Figure 1-1 shows some of the interactions that occur in a forage-livestock system. The farmer brings all these components and their interactions together through management.

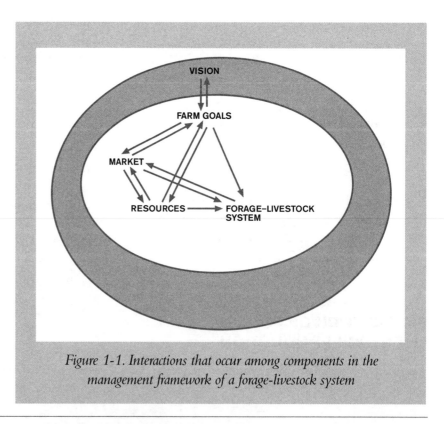

Figure 1-1. Interactions that occur among components in the management framework of a forage-livestock system

PRODUCTION CENTERS

Forage-livestock producers are ecosystem managers. Over the years, we have found that production systems that work with ecological systems are more sustainable than those that work against them. Examples include integrated pest management (IPM) systems using beneficial insects to reduce the need for insecticides and plant breeding to develop disease- and insect-resistant plant varieties.

The goal of commercial forage-livestock production is to harvest solar energy and convert it into net cash income. A forage-livestock system can be visualized as four production centers that parallel this flow of energy (figure 1-2). Management of these production centers can be organized in three domains (forage, livestock, and marketing) that span the production centers. Managing a forage-livestock system entails managing the ecological cycling that occurs in soil fertility, plant growth, and animal production.

Nature uses these basic cycles to generate production in any ecosystem.

Forage production involves managing the genetic ability of adapted plants to harvest solar energy and nutrient and water cycles to optimize forage growth and quality.

Forage utilization is managing the timing and intensity of forage harvest to control forage growth and quality, utilization efficiency, and animal performance.

Animal production involves managing the genetic ability and health of adapted animals to convert the utilized forage into an animal product to be marketed.

Marketing, discussed above, appears as a common theme in all perspectives of a forage-livestock system.

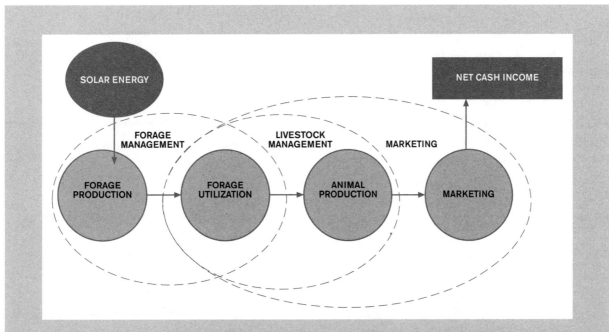

Figure 1-2. The four production centers and three management domains in a forage-livestock system that parallel the flow of energy through the system, and should result in positive net cash income

SUSTAINABILITY

A sustainable farm must achieve social, economic, and environmental goals. An environmentally sound farm that fails to provide an acceptable standard of living or quality of life will not last long. Likewise, farms that are profitable for the short term at the expense of the environment will have a limited lifespan. In general, sustainable farmers focus on two strategies: reducing chemical dependence and using ecological practices (18). Sustainability is a process. Today's sustainable practices will give way to more sustainable practices in tomorrow's economy, market, and environment.

Social Sustainability

Management-intensive grazing has proven to be a way for farmers to use the existing, underutilized pasture resource for feeding livestock to cut production costs and reduce labor needs, thereby gaining the money, time, and energy needed to improve their family's quality of life (1, 19). Pasture-based dairy farmers work less (12) and often have a higher profit margin per cow (28) than confinement farmers, potentially enabling them to have a better quality of life. Pasture-based dairy farmers are more optimistic about their farming prospects than confinement farmers (12).

Sustainability's social benefits arise in various ways. Feeding livestock on well-managed pasture often improves a farm family's quality of life by reducing stress and labor. As livestock consume more forage from pasture, farmers spend less on purchased feed, fuel, fertilizer, and pesticides. The purchase of these products drains money from rural communities, contributing to their impoverishment. Pasture-based farmers have more disposable income to spend on products and services provided by the local community. A ripple effect generates $1.50–2.80 for each $1 spent within the local economy (7). Because more farms may remain in business due to higher profitability and less work (12),

more farm children may remain in farming because they perceive it as a desirable occupation. This cycle helps to rejuvenate rural communities.

Economic Sustainability

Livestock producers are developing innovative production systems. Many are not enlarging their farms in an attempt to increase profitability by spreading production costs over more units of production; rather, they are reducing capital investments and variable input costs, thus lowering unit production costs (3, 12). The idea is to make better use of pastureland through management-intensive grazing, which emphasizes the effective management of forage and livestock to achieve desired goals. A major component of this strategy is controlling the frequency and intensity of grazing to improve livestock feeding on pasture. Many of the basic principles were identified in France (26), based on European and American research, and later refined by farmers and researchers in New Zealand and elsewhere (15). Although management-intensive grazing has been a common practice in other countries for 35 years (3), it is an innovative, alternative technology for many American farmers. It is a technology that requires management skills rather than high capital investment in equipment, buildings, and other variable costs, and it is suitable for any size farm.

Better use of the pasture resource often helps solve the problems of low profitability and excessive workload, especially for dairy farmers. Feeding livestock on pasture can cost only one-sixth as much as feeding them in confinement (23). Compared with confinement, the benefits of pasture feeding may include:

*   lower labor and machinery costs used in cropping, harvesting, storing feed, and manure handling;

- decreased volume and costs of grain and concentrate supplements fed;

- sale of surplus feed or less purchased feed;

- lower veterinary costs because animals on pasture are healthier and have better reproductive performance;

- longer productive life for livestock (lower cull rates); and

- potential premium price for high-quality milk (with low somatic cell count) from pastured cows (5, 12, 17, 22, 24).

As a result of these benefits, 22 Vermont pasture-based dairy farmers achieved an average of $600 net cash income per cow during 1994–1995. In contrast, the top 25% most profitable Vermont farms using Agrifax accounting (24 farms, most likely confinement) obtained an average of only $450 net cash income per cow (28).

Environmental Sustainability

Environmentally sustainable agriculture works with the ecological processes that sustain nature. In nature, nothing is wasted because the outputs of every process constitute the inputs to other processes. For example, green leaves cover the soil throughout the growing season, capturing solar energy by photosynthesis while protecting the soil. Carbon, captured via photosynthesis from carbon dioxide in the air, fixes the solar energy and moves it sequentially from producer, to consumer, to decomposer. Each siphons off its share of energy before passing the remainder on to the next level. Ultimately, the carbon returns to the atmosphere as carbon dioxide, completing one in an endless series of cycles of life, death, decomposition, and new life. Mineral nutrients and water follow their own cycles, parallel to the flow of carbon. When properly managed, these cyclical, natural processes conserve and enhance soil, water, and air quality, and can promote functional biodiversity—all of which are central to environmental sustainability.

Well-managed pasture systems need to work with the ecological processes that maintain sustainability in nature. In contrast, many livestock production systems that involve specialized crop production to support livestock feeding in confinement depart fundamentally from the economy of nature.

BENEFITS OF PASTURE-BASED SYSTEMS

Benefits of pasture-based systems include soil conservation and health, nutrient cycling, reduced biocide dependence, and maintenance of water quality.

Soil Conservation and Health

Year-round ground cover provided by perennial pastures protects and improves the soil. Both living and dead plants intercept rainfall and reduce soil erosion. Rainfall is particularly damaging to bare soil because it dislodges soil particles, causing crusting that impedes water infiltration into the soil and washes soil into streams. Growing annual crops for livestock feed incurs erosion risks that are largely absent in grass-based systems.

A comparison of soil erosion potential was made between pasture-based and in-barn feeding systems in New York State (22). Two levels of milk production were used (16,402 versus 18,370 pounds milk per cow) with each scenario designed to produce 11.3 billion pounds of milk, the market volume expected in 1998. It was estimated that more acreage would be needed to support the pasture option, particularly at the higher per-cow output. However, the type of crops grown would change as a larger proportion of soil-conserving crops was included in the pasture option. Estimated potential soil loss for the state was 33% and 27% less in the pasture option under the moderate and high milk production scenarios, respectively. Retention of both the soil and soil-bound nutrients on the farm and out of the water is of particular importance in a region with a high population of downstream users.

Plants and plant parts left on the ground serve as mulch, which retards evaporative loss of soil moisture. The high specific-heat capacity of water reduces soil temperature extremes on a daily and annual basis, which helps plant roots, soil fauna, and beneficial microbial activity.

Perennial grasslands not only reduce soil erosion but also form new soil and increase soil organic matter (2). Unlike annuals, which grow for just a few months of the year, perennials grow and contribute to soil organic matter from early spring to late fall. Holding land in perennial pasture promotes the accumulation of organic matter in the soil. Among the benefits of soil organic matter are a vigorous soil biotic community and improved aeration due to soil macro- and microfauna activity and root growth and dieback. Macropore continuity from the surface to the subsoil, which is prevented by annual tillage, enhances water infiltration, reducing runoff and the risk of erosion and surface water contamination.

Increases in soil organic matter can potentially decrease atmospheric carbon dioxide and the risk of global warming (10, 13). Grasslands in the Conservation Reserve Program (CRP) are gaining an average of 1,000 pounds of carbon per acre per year. These results suggest that the 45 million acres enrolled in CRP are recovering about 45% of the 84 billion pounds of carbon released annually from U.S. agricultural activities (9). Established pasture soils contain about twice the organic matter of tilled soils. Due to this difference, for each acre converted from tilled crops to permanent pasture, about 78 tons of total carbon dioxide can be removed from the atmosphere and the carbon stored in the soil (22).

Nutrient Cycling

Perennial sods promote efficient nutrient cycling. They maintain active nutrient uptake during much of the year, reducing the tendency for nutrients to move below the rooting zone. In the Northeast, precipitation exceeds evapotranspiration in the spring, fall, and winter, resulting in water moving down through the soil. During these seasons, the risk of leaching is greatest in annual warm-season crops because they are not actively growing. Deep-rooting forages, such as alfalfa and birdsfoot trefoil, can intercept and take up nutrients from lower levels in the soil profile, recycling nutrients that would be lost from the system.

Manure is a valued source of nutrients on a grassland farm, but it has become a major liability on many confined-feeding operations. When a significant proportion of the ration comes from imported feeds, application of manure to a limited land base can lead to excessive nutrient levels in soils. On pasture, livestock distribute the manure themselves. Still, when imported feeds constitute a significant portion of the pastured livestock ration, nutrient accumulation and pollution are likely. A well-designed grazing system avoids manure accumulation and encroachment on surface waters, effectively internalizing nutrient flows. In pragmatic terms, downstream and adjacent neighbors will have much less to complain about with a grass-based system compared to a confinement-based system.

Organic matter accumulated in pasture soils creates a reservoir for nutrient accumulation and release through the action of the soil's biotic community. Soil temperature regulates biotic activity and synchronizes nutrient release with plant root activity and uptake, thus minimizing the risk of leaching loss and groundwater contamination.

On sloping land, grassland slows the overland flow of water and sediment, and hence reduces loss of both dissolved and soil-bound nutrients. Therefore, perennial pastures constitute a lower risk of surface water pollution than tilled land.

Pastures can pose some environmental risks. Pasture may be vulnerable to off-site movement of soluble potassium, to denitrification, and to nitrogen leaching, particularly with high stock densities on highly fertilized swards. Sediment suspension and delivery, and stream bank destabilization are also issues when livestock have access to flowing watercourses.

Reduced Biocide Dependence

Well-managed pastures seldom need any insecticide or herbicide. Crops grown for confinement feeding, such as corn and soybeans, depend on insecticides and herbicides. As a result, corn and soybeans, which account for only a third of the crop acreage in Ontario, for example, receive 84% of the herbicides applied to field crops. These crops are also among the heavier users of insecticides. Thus, dependence on corn and soybeans for livestock feed promotes the use of biocides, with their attendant risks (11, 14, 21, 20, 4, 8). At present, 70–90% of the grain grown in North America, and 40% worldwide, is used to feed livestock. By replacing annual grains with perennial pastures, dependence on and exposure to biocides are reduced.

Maintenance of Water Quality

Surface water and groundwater quality that has been degraded by soil sediments, nitrogen, phosphorus, or pesticide runoff and leaching from cultivated cropland can be improved by substituting permanent pasture for corn and other tilled crops. Farmers feeding livestock on intensively managed permanent pasture purchase less fertilizer, pesticides, and supplements. Consequently, fewer pollutants enter and leave permanent pasture-based farms (27). For example, we estimated that for each dairy cow fed for six

months on intensively managed pasture in Vermont, 37 pounds less potassium, 15 pounds less nitrogen, and 0.7 pound less herbicide would enter a farm than for a confined animal over the same period, due to reduced supplementation and tilled cropping activities (16). Society is also increasingly unwilling to accept the downstream implications of high-density livestock operations, particularly those relating to manure. Manure contamination of surface and well water exposes humans to risk of infection by *Cryptosporidia,* pathogenic coliforms, *Pfiesteria,* and other pathogens.

In sum, grass-based livestock production systems are intrinsically endowed with many of the characteristics that sustain nature. Year-round ground cover, with active nutrient cycling from early spring to late fall, safeguards natural resources and minimizes off-site effects. Stocking animals at the rate the land will bear, rather than relying on imported feeds, prevents nutrient pollution and associated human health issues. Risks to humans, livestock, and wildlife are reduced by the virtual absence of biocides on pasture. Grass farming is an approach that meshes well with society's demand for safe and healthful food produced in a way that enhances and sustains the environment.

The strength of America's agriculture is its diversity, which is based on a variety of climates, soils, markets, and farmers. The array of different production systems is the result of many farmers responding to local climates, soils, and markets based on their viewpoints and skills, and on how they can best achieve their goals using the resources at hand. Agriculture has had its economic ups and downs over the years, but food production has been more than adequate. The past stability in American agriculture is an example of the ecological principle of stability being founded on diversity, whereas production volatility is founded on simplicity.

THE SCIENCE AND THE ART OF PASTURE-BASED SYSTEMS

The purpose of this book is to provide fundamental and essential information a manager needs to manage and market a goal-oriented forage-livestock system. The chapters are organized by a production-center approach that parallels the flow of energy through the system.

There is both science and art involved in forage-livestock system development and management. The science includes the ecological principles relating to soils, water, plants, and animals, and describes how they respond to changes in physical and management inputs. The art includes the ability to manage these responses and their interactions to maintain a productive, sustainable system.

Science-based principles can be provided by books, fact sheets, and educational meetings. Application of these principles and the critical evaluation of their results can help the manager develop the artistic skills needed to apply the principles over a varied landscape.

Pasture-based livestock systems need to be managed for an economic optimum, not a maximum product output, because there are social, economic, and environmental parameters that do not lend themselves well to maximization.

As farmers develop their own forage-livestock systems, it is important that they read and listen to a number of viewpoints so that they can see the potential of alternative management systems and find components that best help meet their particular goals.

CHAPTER 2
Vision, Mission, and Goals

Lisa Holden and David Grusenmeyer

"Greatness is not in where we stand, but in what direction we are moving. We must sail sometimes with the wind and sometimes against it— but sail we must, and not drift, nor lie at anchor." *— Oliver Wendell Holmes*

There is a saying often quoted in times of change that "a bend in the road is not the end of the road, unless you fail to make the turn." One thing that you can count on in a pasture-based feeding system is change. The weather will change, your swards will change, and your management skills will change.

This chapter looks at vision and mission statements and setting goals—those things that don't change as frequently as the weather and that act as a guide for operation. Having written vision and mission statements and written goals gives direction to your operation so that day-to-day decisions lead toward the business goals.

THE VISION STATEMENT

What Is a Vision Statement?

Close your eyes and picture the best pasture, with the straightest fence and the healthiest animals. Do you have that picture clear in your mind? That's vision—seeing something that does not now exist as you would like it to be in the future.

Now, do the same thing for your whole business. A vision is a statement, or picture, of the future that you're working to create—what you want your business to become.

Without a vision of where you're going, how can you develop a plan to get there and how will you know when you've arrived? Without a vision of

where we would like to be, we can continue hiking various trails through life, climbing mountain after mountain, only to discover each time that we've arrived somewhere we really don't want to be. Nothing worthwhile was ever created without a vision. It guides us, gives us direction and purpose, and can serve as a powerful motivator for us and those around us. To truly guide and motivate, a vision must be:

- aligned with the core values of both the individuals and the farm business; and
- effectively communicated to and accepted by everyone involved in the farm.

The more precise and detailed you can be in writing a description of your vision of the future, the easier it will be to communicate it to others and gain their commitment to it, and the more likely you will be to achieve it.

How Is a Vision Statement Developed?

Developing a vision statement takes time, energy, and lots of hard work. It involves looking at your core values and thinking strategically about the future. Developing vision also involves other people: your family, your farm advisors, your employees.

Core Values

Core values are those principles and standards at the very center of our spirits, from which we will not budge or stray. Core values are extremely stable and change only very slowly over long periods of time. Core values are like the corner posts in a fence: they hold everything together and determine the shape and the strength of the final product that we see. Your core values do the same for you; they shape your attitudes and behaviors.

For some people, identifying and communicating personal core values can be a difficult task. Core values are so close to the center of who we are that they tend to be very protected and not shared with others until a significant personal relationship has been established. Because these values are so central to what's important to us, it is all the more important that we think about them first as a basis for establishing a sound and meaningful mission, a vision, and goals in both life and business.

Strategic View

Having a strategic view means looking beyond the pasture farthest from the barn. It means learning about what is happening in your community and your industry. What are the opportunities for growth? (This involves not just animal numbers, but personal learning experiences as well as production and profit.) Are there "bends in the road" on the horizon that your operation will be facing? Are there exciting new trends that will help sustain your business? Taking time to think strategically about new challenges and opportunities will help you use your core values to develop your vision.

It is easy to focus on what we can see, hear, smell, and touch. The future is elusive, but developing and realizing that picture-perfect vision starts with stretching your mind and looking ahead.

Involving Others

Developing your vision statement for your farm should involve everyone in the operation who is involved in making decisions. Additionally, asking for advice from your farm advisors can help you clarify your vision and mission statements. You need to share your vision with others to help

refine and sharpen it for yourself. Talking with family and nonfamily employees and posting a final copy of the vision, mission, and goals for your operation will help others understand the direction in which the farm is moving and what is most important.

You can't expect that just because you develop mission and vision statements and post printed copies everyone will immediately accept and work toward achieving them. You first need to "walk the talk" and be totally committed to them yourself. Then discuss them with your employees and consultants. You may need to discuss them several times before they will believe you're really serious and begin to internalize the statements.

MISSION STATEMENTS

What Is a Farm Mission Statement?

A vision statement is a picture of what you want your farm to look like in the future; a mission statement is a snapshot of what the farm is and does right now. The mission is a broad statement of business scope, purpose, and operation that distinguishes your farm from others. The next time you are visiting a business (restaurant, retail store, bank, whatever) take a look around and see if there is a mission statement posted somewhere on the wall. That mission statement tells everyone who enters the business something about what its purpose is and what is important. Your mission statement should do the same thing for those people who visit your farm.

How Do I Write a Farm Mission Statement?

Asking the following questions can help you write a farm mission statement:

*   What am I in business to do? What is my purpose?

- How is my business unique from other businesses like mine?

- What do I value in a farm business?

Values, beliefs, and a mission are formulated by people, and a business cannot have them apart from the people who make up that business. Therefore, especially for small, closely held businesses, it's important that each person in the business write his own personal mission statement first. Then everyone should come together to develop a mission statement for the farm.

A farm business mission statement reflects the core values and beliefs of the individuals who lead the business. To the extent that there are large differences between a farm mission and a personal mission, or between farm business values and personal core values, there will be frustration and dissatisfaction between individuals within the business.

In addition to giving structure and direction to an individual or business, well-written mission statements are excellent tools to inform others about what's important to you and how you operate your business. Worksheet 1 (page 12) will help you develop your farm mission statement.

The sample mission statements below communicate very different notions about what's important on these farms and also give some indication that day-to-day business may be conducted differently as a result. Any mission statement that concisely represents truth and reality about the farm is a good mission statement. Likewise, any statement that doesn't honestly and accurately represent the values and beliefs of the individuals who lead the farm is a poor mission statement, regardless of what it says or how good it sounds. If excellence is a stated value or the pursuit of excellence is a stated mission, yet an average, industry-standard, or legal requirement is "good enough," then what is the real commitment to excellence?

Mission statements serve to inform employees, friends, neighbors, and those in agribusinesses about what's important to you and your business. They also serve as anchors and guideposts for both strategic and operational decision-making on the farm.

Sample Mission Statements

"At Maple Hill Farm, our priorities are God, family (people), business. We strive to be a place where people (our most valuable asset) have the opportunity to grow spiritually, personally, intellectually, and financially. By putting God first and people second, our personal success and success as a business is guaranteed."

"Our goal is to produce large volumes of high-quality milk as economically as possible, in order to provide an adequate standard of living for the owners and our employees."

"Green Acres Farm is a grass-based dairy that provides the public with safe, wholesome milk from healthy cows. Our family-owned and -operated dairy is a safe and friendly place to work."

1. What am I in business to do? What products do I produce for which consumers?

_____

_____

_____

_____

2. What are the unique qualities that separate my business from others?
   What strengths does my farm business have?

_____

_____

_____

_____

3. What values provide the foundation for my farm business?

_____

_____

_____

_____

GOAL SETTING

Why Is It Important to Set Goals?

Goals are very powerful tools. Setting goals helps to enhance communication among people working on the farm. Having written goals helps you focus. Goals are important because they direct behavior. If you need to prove that to yourself, try this simple exercise. Set a goal for yourself for the next week. Write that goal on three pieces of paper. Hang one piece on your bathroom mirror at eye level. Hang the second piece on the refrigerator. Hang the third piece on the dashboard of your car or truck. At the end of the week, have you accomplished the goal? Goals that are written down and visible (e.g., hanging on the refrigerator) are much more likely to be achieved than those that we keep in our heads.

Mission and vision statements, although frequently short, are broad, encompassing, and far-reaching. They can often seem overwhelming and impossible to achieve. The metaphors, "How do you eat an elephant? One bite at a time," and "A journey of a thousand miles begins with the first step," apply to achieving a mission and vision.

Goals create the bite-size pieces, the road map, and manageable stepping-stones needed to achieve the mission, to make the vision a reality, and to navigate the course you have set for your business.

How Are Goals Set?

Begin by thinking about the different areas of the farm. Each area, as well as the overall operation, should have a set of specific goals. To be effective, goals must be written. If they aren't in writing, they're merely ideas with no real power or conviction behind them. Clearly and specifically written, they also eliminate confusion and misunderstanding. Worksheet 2 (page 14) will help you set goals for your farm business.

Use the SMART format in formulating goals for your farm. Goals should be:

Specific

Measurable

Attainable

Rewarding

Time bound

SAMPLE GOAL: Increase income over feed cost by 20% in the next year.

The "S" for *specific* is the first part of writing the goal. The stated goal to increase income over feed costs by 20% is more specific and thus easier to understand than simply "increase income." Likewise, "Improve legume content in swards for north pastures by 10%," is spelled out clearly enough for everyone to know and understand the goal.

If you can't measure it, you can't manage it, so the "M" for *measurable* is another important part of the goal. Without a measurement, how would you know if you have achieved or exceeded your goal? Are you able to measure income over feed costs?

Yes. Could you determine if income over feed costs changed by 20% from one year to the next? Yes.

The "A" for *attainable* means that goals should be challenging but realistic. Reaching goals is a positive pat on the back for everyone involved in the farm. If goals are so far out of reach that you hardly ever achieve them, everyone gets discouraged. If the goals are too easily reached, they lose their power to motivate. Every situation is different. A 20% improvement as stated in the example is attainable for some operations but perhaps not for others.

The "R" for *rewarding* is the motivational force of the goal. Don't set a goal because you read about it in a book or because it's one your neighbor has. Set goals that provide a reward to your farm, and celebrate receiving those rewards when the goals are attained. Rewards need not be monetary. A sense of satisfaction, pride in accomplishment, and recognition of achievements are all rewards.

The "T" for *time bound* is probably the most important part of the goal. This part lets you know whether or not you have succeeded within the necessary time frame. It also helps you determine what is attainable. The 20% increase in the example may be within easy reach in a year, but not in two months.

Using Goals as Stepping-Stones

Goals are like street names on a road map — each one achieved (or each street passed) brings you one step closer to your vision or destination. Without goals, vision and mission may seem unrealistic.

No matter what plants are being grazed by what species of animals, setting and achieving positive goals benefits the system, the people, and the overall operation. Take that first step by using the worksheets in this chapter to begin setting your goals.

Worksheet 2: Setting "SMART" Goals   Today's Date: _____

GRAZING SYSTEM

Specific goal _____

_____

Who is responsible _____

Completion date _____

ANIMALS

Specific goal _____

_____

Who is responsible _____

Completion date _____

MY FAMILY, MY EMPLOYEES, AND MYSELF

Specific goal _____

_____

Who is responsible _____

Completion date _____

OVERALL GOALS FOR MY FARM BUSINESS

Specific goal _____

_____

Who is responsible _____

Completion date _____

Remember that goals should be: Specific, Measurable, Attainable, Rewarding, and Time Bound

Researchers studying the effects of goals as part of a company's overall performance management process found that the level of performance is highest when:

- goals are clearly stated and contain specific objectives,
- goals are challenging but not unreasonable,
- employees accept their goals with a true sense of ownership,
- employees participate in setting and reviewing their goals.

As time goes on and goals are achieved, or conditions and situations change, it's important to reevaluate and establish new goals. Failure to periodically set new or more challenging goals can lead to stagnation in the business and boredom among employees. Finally, as goals are achieved, providing positive feedback and rewards for yourself, family members, and employees is critical to maintaining enthusiasm and continued progress.

CONCLUSION

Bob Milligan of Cornell University uses a great goal analogy involving a group of people carpooling together.

It would be very difficult for everyone in a carpool to make a decision on whether to turn right, turn left, or go straight at the next intersection if each was headed for a different destination. If they're all going to the same place, they may have different ideas on which way to turn and exactly how to get to where they're going. One may want to take the scenic route, another knows about road construction that should be avoided, a third may want to take a shortcut and arrive early, and a fourth may want to run an errand along the way. But they should be able to reach consensus on the route to take based on information provided by each rider, since their endpoint is the same.

For your grazing system and your overall operation, the time you spend planning and developing your vision, mission, and goals will be just as important as the time that you spend tending to pastures and animals—maybe even more important because of the direction and guidance they provide for planning and decision making.

Developing a shared mission, vision, and goals is the first step in laying a foundation for making strategic and tactical decisions that will move your farm forward. Having them in place won't eliminate arguments and disagreements, but at least the disagreements will be about how best to get to the same endpoint as opposed to heading in opposite directions.

So before that first blade of grass turns green in the spring, stop and take time to think about the vision, mission, and goals for your farm. They are just as vital to your grazing system and your business as the soil, water, plants, and animals.

CHAPTER 3
Resource Inventories in Farm Planning

Darrell L. Emmick, Julian Drelich, Jr., and James S. Hill

Systematic farm planning activities require having access to information and knowledge of the natural and physical resources existing on a particular farm or ranch. Therefore, conducting a resource inventory should be one of the very first steps in the development of an economically viable and environmentally sustainable pasture-based livestock production system. Attempting to engage in a livestock production-based business without first assessing the natural and physical resource base and determining its capabilities and limits simply dooms the enterprise to a high probability of failure.

The U.S. Department of Agriculture, Natural Resources Conservation Service (USDA, NRCS) defines a resource inventory as the collection, assemblage, interpretation, and analysis of natural and physical resources for use in farm planning activities or for other resource management purposes (1). A resource inventory identifies not only what resources are available on a farm or ranch but also establishes the baseline or benchmark condition of each of the resources identified.

The primary natural resources to be inventoried are land, soil, water, air, plants, and animals. Physical resources include human resources, financial resources, facilities, and equipment. Obtaining information on each of these components is necessary in the development of an overall farm plan; however, the amount and scope of information required for each component will depend on the nature of the enterprise and the significance of the component to the success or failure of the operation.

PRESCRIBED GRAZING MANAGEMENT: A SYSTEMATIC PLANNING STRATEGY

Prescribed grazing management (PGM) is defined by the NRCS as the controlled harvest of vegetation with grazing or browsing animals managed with the intent to achieve a specified objective (2). PGM is not a grazing system; it is a systematic planning strategy that blends the goals and objectives of a particular producer with the natural and physical resources on a particular farm or ranch to develop an economically, environmentally, and ecologically sound grazing management plan.

PGM is not a "one size fits all" approach to grazing management planning. The PGM planning procedure is an information-based strategy that utilizes location-specific resource data and producer-provided information in the development of individual plans. Because of this approach, each farm or ranch will have a unique plan. Each PGM plan represents a site-specific application of principles and practices that have been selected to address the resource management needs of a particular farm or ranch, as well as the needs and objectives of the producer.

The following nine-step process is an application of the NRCS conservation planning process as presented in the NRCS *National Planning Procedures Handbook* (1). This is a progressive problem-solving and management procedure that provides landowners with a systematic way to look at and evaluate the resources on their farms and to plan an effective management strategy for those resources (1). The approach emphasizes desired future conditions. For this process to be effective producers must have a clear understanding of their goals and objectives.

Identifying Goals and Objectives

To begin this section, we ask the question, If you do not know where you are going, how will you know when you get there? Setting goals and objectives is something that should be done before—not after—you spend your money. Unfortunately, many situations arise in which people acquire livestock, land, or both without really having a goal or objective in mind; as a result, they simply turn potential assets into financial liabilities.

Planning is often viewed as a means to achieve an end. A process, however, is more a series of actions or operations leading to a desired condition or outcome. The systematic planning of a farm or ranch operation represents a *process* designed to achieve a predetermined goal (table 3-1). Goals drive the planning process and ultimately the selection and subsequent adoption of specific management practices.

Regardless of the goal, it should be written down in a manner that is clear, concise, and easily understood by all of the parties involved.

To ensure that you are not wasting time, money, and effort on practices, products, or implementation strategies that are not required to attain a particular goal, make sure that you know what it is you want to do before you get started and be as specific as you can.

Some examples of specific, concise goals are:

- Graze 50 dairy replacement heifers on 50 acres of former hayland.

- Increase the harvest efficiency by 20% on an existing pasture.

- Use 30 goats to clear brush from 40 acres of abandoned farmland.

- Develop a 100-ewe grass-based milking sheep operation on a recently acquired farm.

- Increase the size of a beef cow-calf operation from 50 to 75 cow-calf pairs.

- Net enough money from grazing dairy replacement heifers to pay the farm mortgage.

Trying to develop a management strategy to facilitate goals that are not clearly stated is much like being lost in a snowstorm. A lot of time and energy is spent traveling in circles and getting nowhere.

Table 3-1: Sample Goals and Corresponding Issues for Each	
GOAL	**ISSUES TO CONSIDER**
Clear brush with livestock	What kind of animals? How many?
Make best use of existing pastures	Hay, graze, combination of both? Is the objective pounds per animal or production per acre?
Realize noneconomic goals, such as the pleasures of watching animals on the farm.	Time, labor, family life, financial?

Identifying Problems, Concerns, and Opportunities

Once the goals have been determined, the next step is to identify potential problems or concerns that could stand in the way of achieving them. In some situations, what is initially identified as a problem may actually turn out to be an opportunity. Prospective opportunities should also be identified.

In some situations, an identified problem or concern might be relatively minor, and thus easily solved. For example, having no fence around hayland you want to graze is remedied by simply building a perimeter fence. In other cases, a problem or concern might pose a greater challenge and thus require a more involved solution. Starting a new business on a recently acquired farm requires the complete development of an operational plan, which in turn may indicate the need for everything from constructing a fence to building a new barn and milking or handling facility to addressing species composition in pasture and hayfields. Keep in mind that you may not be in a position to recognize if you have problems, concerns, or opportunities until your goals have been clearly identified. Each farm, livestock species, and landowner represents a unique situation, and what might be a concern on one farm may not be on another. For example, to dairy producers brush-covered pastures are a concern. However, to meat-goat producers, brush-covered pastures may be an ideal forage base. This is an example of a potential problem that might ultimately be an opportunity to enhance income.

If you have already acquired land, it must be evaluated to determine the kind, amount, and seasonal availability of forage or browse it can produce and the physical constraints to its use by livestock. Physical constraints include availability of water, distance from handling facilities, steepness of slope, moisture conditions of the soil, and the presence of boulders, rocks, or other physical impediments to grazing animals. Not all land is capable of producing the same amount or kind of forage and/or browse due to the inherent characteristics of the soil type, and not all land is equally well suited for use by all kinds and classes of livestock.

In some situations, you might already have livestock, but not the right kind, number, or class of animals to do a particular job. For example, you might have 50 acres of brush-covered pasture and 75 lactating dairy cows, and your objective is to clear brush and maintain milk production in the process. Brush is not considered a dairy-quality feed, and even if the forage species in the pasture were of adequate quality, chances are 50 acres of mostly brush would not provide enough feed for 75 dairy cows for very long. Clearing brush is one goal, maintaining milk production is another, and the two are not compatible. If brush clearing is your main goal, then goats, beef cattle, sheep, or a bulldozer and brush hog make better choices. If maintaining milk production on pasture is your main goal, then the major problem or concern will be improving and maintaining the forage base to a quality that will support high-producing milk cows.

Inventorying the Resource Base

Once the goals for an operation have been identified, along with potential problems, concerns, and opportunities, it is time to inventory the existing resource base.

Inventorying your resources helps you develop an understanding of what resources you have available for use, what condition they are in, and how well they actually fit with your stated goals. The resource inventory is conducted at the beginning of the planning process. It identifies the resources in the benchmark, or existing, condition. Benchmark conditions are then used to evaluate the impacts and effects of the planning activities implemented.

A complete resource inventory will include information on land, soil, water, air, plants, and animals, as well as human resources such as labor and management, equipment, facilities, and finances.

Soil and Land Resources

The first step in inventorying your land base is to obtain soil maps, topographic maps, and aerial photographs from your local USDA Service Center. Other maps that may be useful include floodplain areas, stream classification, and state and federal wetland maps. Employees of the NRCS, the Farm Service Agency (FSA), or your county Soil and Water Conservation District (SWCD) routinely use these photos and maps for farm planning and other program activities, and they can assist you. You should also contact the Underground Facilities Protective Organization or your local equivalent to find out what, if any, underground pipelines, electric lines, or communication cables exist on your property, and where they are located.

With maps in hand, delineate the boundaries of your property and determine how many acres you have available for grazing, hay production, crop production, and other uses. Next, using a soil map and interpretation data obtained from your county soil survey, identify, field by field, the major soil types and determine their uses and yield capabilities, which are affected by how steep the land is and how wet or stony the soil is, among other factors.

Analyzing maps and photographs is a good place to start. However, make sure that you complete your analysis by getting out on the land and walking it. Soil erosion, winter-killed forages, and a new beaver pond where pasture once existed cannot be observed from the office or kitchen table.

Walking the land also provides an opportunity to take soil samples for fertility analysis. Knowing what kind and how many acres of land you have available and what the soil capabilities are is the first step in determining what kind and what number of livestock are best suited to your farm.

Livestock

The primary livestock attributes to inventory include the species, number, and class of animal, as well as their combined weight. Animals differ in their nutritional requirements, preference for food items, digestive morphology, body size, agility, and physical ability to negotiate steep slopes and wet soils or to travel long distances in search of food and water. To successfully raise livestock on pasture, you need to be aware of their physical requirements as well as their production attributes and limitations.

Forage

Knowledge of the kind and yield of plants growing in a pasture is important when planning the species and number of animals you should have. Identify grasses, legumes, forbs, and woody species, and determine their seasonal patterns of production as well as current seasonal yields. Seek out information on crop history and pesticide usage. Note problems with stand health and vigor, weeds, toxic plants, and the like. The more detailed the information collected, the more accurate will be the benchmark condition identified.

Water Supply

An adequate supply of clean water is as crucial to grazing animals as food, shelter, and living space. Make sure that the water source is not contaminated and will not be contaminated by your livestock. Make sure there is adequate volume to meet the demands of your herd or flock for as long a time as they will be on pasture. Make sure the water distribution network is adequate and that the pipes, fittings, and valves are all in good working order. Lastly, evaluate the water dispensing site or location where the livestock will have access to the water to ensure that manure and urine contamination will not degrade either the water source or the physical location.

During the resource inventory is also a good time to locate potential sites for water development. Note on an aerial photo any location that may be developed as a pond, spring, or other livestock water source.

Air

Air must be inventoried for movement and temperature. Moving air (wind direction and speed) can cause problems both on and off the farm. Note the prevailing wind direction and its velocity during different times of the year. Off-site and downwind neighbors may be affected by odors, dust, and chemicals emanating from your operation. Conversely, you and your livestock may be affected by the activities of upwind neighbors. In winter, wind chill can be a problem for unprotected livestock and people. In spring, wind can offer relief from flying insects, and in summer, the flow of air can help reduce heat stress.

Livestock Access Routes

Starting at the barn, handling facility, or watering site, walk the paths, lanes, or routes that livestock will need to travel to and from pasture and between pastures. Chances are that anything that would deter or impede you from walking through a particular location or to a specific site would also deter your livestock. Look for and identify problems and concerns such as mud or manure-filled barnyards, wet muddy lanes, steep slopes, or sharp rocks or stones in the lane. Look for problems with stream crossings, wet spots, or other areas where water may pond and cause mud holes to form. Ensure that livestock access routes are safe, easily traveled in all weather conditions likely to be incurred during the period of use, and positioned in environmentally acceptable locations.

Infrastructure, Machinery, and Financial and Human Resources

Infrastructure includes such things as buildings and facilities, lanes, fencing, and water systems. Machinery includes tractors, all-terrain vehicles, and associated implements. Financial resources include your readily available working capital and your access to credit. Human resources are all of the people you can count on to be available when they are needed. Make sure that you know what each person's strengths and weaknesses are and the skills they possess.

Because each livestock enterprise is unique, each will have a different kind, amount, size, and number of these components. However, all need to be evaluated in their current condition in relation to the function they are expected to provide or the task they are required to perform.

In some situations, machinery may be eliminated, downsized, or replaced with a different type of machine—for example, replacing a bigger tractor with a smaller one, replacing a smaller tractor with a four-wheeler, or eliminating a line of equipment. In other situations, a 60-horsepower tractor with a mower-conditioner and a round bailer may have to be purchased.

Current Management System

Identify the key components and practices in the current management system. Describe what it is you do, how you do it, and when. In other words, evaluate what it is you are currently doing in light of what you are trying to accomplish. Most individuals are not good at self-evaluation. As a result, this would be an excellent opportunity to involve the trained personnel from your local NRCS, SWCD, or cooperative extension office.

Wildlife, Rare and Endangered Species, and Cultural Resources

A thorough resource inventory will include not only the specific resources related to production, but also the resources that are ancillary or incidental to the operation.

Although they are often overlooked, society currently places a high value on wildlife, rare and endangered species, and cultural resources. In some cases, there will be value in preserving a piece of history for future generations to enjoy or in maintaining biodiversity. In other situations, there may be value of a financial nature, such as leasing land for hunting, fishing, or other recreational use.

Regardless of their perceived value, these resources should be acknowledged by landowners and operators as a part of the inventory to ensure that their future use is not impaired or lost simply due to oversight.

Analyzing the Resource Data

Once the resource inventory is completed, it is time to study the information collected and evaluate it in reference to your stated goals. Ask yourself the hard questions, such as:

- Do I have enough money, desire, and human resources to do what I want to do?

- Are the facilities and infrastructure adequate for the species, number, and class of livestock intended?

- Is the forage supply on my farm or ranch adequate to meet the demand of the number of head of livestock I am planning?

- Are the current livestock the right kind, number, and class to meet my goals?

- Do the kind and yield of plants in the forage base complement the nutritional requirements of my livestock?

- Is the land base appropriate for the kind, number, and class of livestock?

- Is the water supply adequate for all seasons of the year?

- Are the pasture access routes acceptable both environmentally and to the livestock?

- Does my current management system facilitate my goals or is it in need of an overhaul?

Be honest with yourself, as this information represents existing or benchmark conditions, and once these conditions are evaluated, they may or may not support what it is you think you want to do. This information will also help establish cause and effect relationships and, as a result, provide insight concerning existing and future conditions (1).

With this step completed, you will have the basic information you need to determine whether or not your goals are realistic, given your resource base and its condition. In addition, this information will enable you to formulate and evaluate alternative management strategies to complement, improve, or make more efficient use of your resource base and to make other business management decisions.

Formulating and Evaluating Alternative Solutions

Before solutions to problems can be evaluated, they must first be formulated. Oftentimes there are several different solutions to a single problem or concern, but they are simply overlooked. There is value in bringing in outside help (NRCS, SWCD, cooperative extension, or private consultants) for a brainstorming session, if you have not already done so.

Once the resource data have been evaluated, you will have a good idea of whether or not what you are planning to do is feasible, or perhaps you will know why what you were trying to do will not work. In either event, you will have the information available to formulate and evaluate new alternatives.

Based on the available resource information, you need to consider identifying and developing as many technically sound alternative management options as possible. These alternatives can then be evaluated for feasibility in your operation.

Making Decisions

Once all of the alternatives have been evaluated for feasibility, it is time to decide whether or not you should go ahead with your plan, modify your plan, or scrap the idea and start all over with a different plan. Good decisions are made when they reflect what is possible based on the evaluation of your resources. Again, be honest! Do not let your emotions cloud the truth. While owning a flock of 300 sheep, managing a herd of 200 cow-calf pairs, or grazing dairy cows may be your personal ambition, if the resource base on your farm or ranch is not compatible with what you want to do, you do not stand much chance of success.

Preparing a Plan

Up to this point, the emphasis has been on gathering and analyzing information that will support the management decisions. Now it is time to develop a plan—not in your head but on paper. The plan should include such things as:

- What needs to be done concerning each field, the livestock, the facilities, finances, labor, and so forth?
- When do the management practices and activities identified need to be done?

- How will I implement the practices and find sources of assistance?
- Who will be responsible for what?
- What type of expertise is needed?

In preparing the plan, you must be realistic. Not everything you have planned may be doable in one year or even two. One of the biggest reasons for failure is trying to do too much too soon. Develop a sequential approach based on your highest priority. For example, what should come first, developing a water source, building a fence, or improving soil fertility? Each farm or ranch is unique; therefore, the answers will vary from operation to operation.

This is another good time to ask for assistance from farm-planning specialists, such as the staff at the USDA Service Centers. They have experience in developing plans and recommending alternatives that are economically and environmentally acceptable.

Implementing the Plan

Once a plan has been prepared, it is time to start putting it "on the ground." Plan a logical sequence of installation. Assuming the planning process has been relatively accurate, the implementation phase can be simple—follow the sequence that you developed. Spend as much time as possible in the field looking at and thinking about what you are planning to do. Challenge yourself by asking, What are some reasons the plan won't work? Sometimes you will discover something has been left out in the planning process.

As stated above, it may take several years before the system you implement becomes fully operational. This is especially true if you are starting a new operation and one part of the plan depends on the successful completion of another part of the plan. Even with the best inventory possible, things may change or turn out not to be what was expected. When this happens, it may be necessary to revisit the resource inventory to see what can be done to get the plan back on track. Sometimes the only solution is to change your goals completely. This may involve another resource inventory to see if the new objectives are feasible.

Evaluating Results

A detailed evaluation of all aspects of the operation must be completed. This should be done as you progress through the implementation phase and after everything has been installed and is operational. Hopefully, everything is proceeding just as you had envisioned. But what if it isn't?

Sometimes things do not go as planned. This does not mean that the resource inventory or planning process is unreliable. It might mean that conditions have changed, that data in the resource inventory were not properly evaluated or were overlooked, or that some unexpected variable has surfaced. When this happens, you must decide if the actual results are satisfactory, or if you need to do something different. If you need to make changes, make them one at a time and make them slowly. In most cases natural resources take time to respond to new or different management strategies. Making too many changes too fast generally compounds the problems and makes it even more difficult to sort out cause and effect relationships. If working with natural resources teaches us nothing else, it teaches us that little is certain and everything is subject to change.

CONCLUSION

The efficient production of livestock from pasture begins with an understanding of the farm's natural and physical resources. Without knowing what you have to work with and how these resources will likely respond to various management strategies, you stand little chance of making the technically sound decisions required to run a successful business.

All business enterprises, be they industrial, manufacturing, or agricultural, should start out with a close examination of the resources available. Conducting a comprehensive resource inventory is a valuable tool in determining what's possible. The more complete the inventory, the better the chances of developing a management strategy that will successfully meet resource-based goals and objectives.

The degree to which the farm is successful will depend upon your ability to recognize the opportunities and concerns associated with the available resources and to design a program or plan that will make best use of those resources.

CHAPTER 4
Allocation of Farm Resources

Stephen A. Ford and Wesley N. Musser

Using grazing in livestock systems has been well documented as a profitable alternative to using confined-feeding systems with harvested forage (4). Grazing is more profitable largely because it is cheaper for livestock to harvest grass than it is to harvest grass mechanically and store it for later feeding. Other costs and benefits also contribute to grazing profitability. However, case studies have shown that grazing is not always more profitable than confined feeding (6). The relative profitability of grazing depends on the application of available managerial strengths either to grazing or to confined-feeding activities, as well as on the existing feeding and housing facilities.

Each farm manager must examine the profit potential of alternative production systems in the context of a specific farm. Because adoption or modification of grazing systems often changes many inputs and outputs on the farm, evaluation of grazing is often quite complex.

This chapter discusses the important aspects of farm management economics as applied to various grazing decisions. It is not meant to replace currently available farm management textbooks (2), but applies principles of farm management economics to grazing issues. Readers are encouraged to learn more about farm management economics in other books to improve their understanding of this topic.

MAKING DECISIONS AT THE MARGIN

Most farm management decisions deal with changes to existing operations. Simple comparisons of costs and benefits of a change in practices, an investment, or the use of an input help determine whether or not to implement the change. Because these tend to be marginal decisions that do not affect the primary focus of the farm operation, a complete analysis of costs and returns to the whole farm is unnecessary. The farm manager only needs to look at the costs and benefits that change as a result of the decision being made.

Most marginal decisions are made at the level of units of production. For crops, that means the economics are examined on an acre basis, and for livestock, economics are examined on a per-head basis. This convention is used primarily so that the units of comparison are similar across enterprises and farms. However, for an individual farm, the units to be considered should comprise the entire change. For example, if 30 acres of hay land are being converted to pasture, the analysis should be for the 30 acres, not an individual acre. If the profitability of one acre can be converted to 30 acres simply by multiplying by 30, consideration of the whole production system is not so useful. However, inputs and production practices often change in a nonlinear fashion as numbers of acres and animal units change in production systems. For example, the conversion of 30 acres to pasture could allow the sale of most of the hay-harvesting equipment if this is the only hay on the farm. However, the haying equipment and many costs associated with it would not change if only 1 acre was converted. Therefore, it is recommended that farm managers consider changes in the entire system in their analyses but include only those costs and benefits that are not constant.

Partial Budgeting

Partial budgeting is an analytical framework that assists farm managers in analyzing the marginal effects of a decision. A partial budget is simply a reminder to consider costs and benefits that are not constant when changes occur and to organize them correctly. When a change is made on the farm, the cost of that change includes the direct expenses associated with it and the indirect cost of any lost revenue as a result of the change. Similarly, the benefits of the change include the new revenue resulting from the change and the indirect benefits of any savings resulting from the change. It is useful to organize these changes as follows:

BENEFITS	COSTS
Increased Revenue	Decreased Revenue
Decreased Expenses	Increased Expenses

If the sum of the benefits of the change is greater than its costs, then the change is profitable.

It is important to include all changing costs and benefits in the analysis. Cash costs are fairly straightforward. However, noncash costs must also be included. Depreciation, inventory changes, and opportunity costs are examples of noncash costs and benefits. It follows then that cash flow and profitability are not necessarily the same. Both are necessary for farm financial health, but it is possible to have one without the other. For example, consider a livestock operation that bought calves in the spring, grazed them through the year, but had not sold them at the end of the year. The operation might be profitable when the value of the weight gain of the animals is considered, but cash flows are negative because no sales were made. Alternatively, if the calves were sold for more cash than total cash expenses, cash flow is positive. However, if cash revenue minus costs for the calves is less than depreciation and other noncash expenses, the operation is not profitable. More discussion of cash flow and profitability distinctions will be presented later in this chapter.

An example of marginal analysis using partial budgeting is presented in table 4-1. This analysis is to determine whether to convert a hay field to pasture. Costs of the change include expenses associated with pasture production and fencing. Let's assume that the cows are watered at the barn without any

Table 4-1: Partial Budget of Converting 30 Acres of Hay to Pasture

BENEFITS		COSTS	
Increased Revenue (from pasture value)		**Decreased Revenue** (from lost hay value)	
Pasture value (3 tons of dry matter per acre @ $90/ton for 30 acres)	$8,100	Hay value (4 tons of dry matter per acre @ $90/ton for 30 acres)	$10,800
Decreased Expenses (no longer haying)		**Increased Expenses** (due to pasture)	
Hay labor (cut, rake, bale, haul)	$1,008	Annual fence repairs	$100
Hay fuel (cut, rake, bale, haul)	$237	Pasture labor (180 hours @ $8)	$1,440
Animal health (30 cows @ $10)	$300	Annualized fence establishment costs*	$1,054
Feed (30 cows @ $20)	$600	($4,215 total cost over 5 years)	
Baler annual fixed costs (sold)	$4,500		
Total Benefits	$14,745	**Total Costs**	$13,394
Net Benefit	$1,351		

*A multiyear allocation of the fencing investment accounting for the opportunity cost of money.

added cost. Another cost, however, is the decreased revenue from hay sales. If hay is fed rather than sold, an increased expense is the purchase cost of hay produced from the field, and decreased revenue would be zero. Benefits of this proposed change from hay to pasture include increased revenue from pasture forage sales and/or reduced expense of feed purchases. However, the primary benefit of pasture is the reduction in hay production expenses. If the benefit of the expense savings outweighs the costs associated with pasture, the change would be advisable. Otherwise, the field should stay in hay, or another option should be evaluated.

While partial budgeting is one of the most powerful analytical tools available to the farm manager, it still has several drawbacks. First, each partial budget compares only a single alternative to the current situation. Evaluation of many alternatives requires multiple budgets. Second, the manager must take care to evaluate separately the profitability of all components included in the change. The inclusion of an unprofitable component may be overshadowed by the overall profitability of the change being considered. For example, in the hay field conversion to pasture analysis presented in table 4-1, an unprofitable fertilizer application level on the pasture may not greatly affect the decision to adopt grazing on that field. However, if the decision were close, a fertilizer expense that is higher than recommended may incorrectly indicate that grazing is not advisable in this example. Finally, partial budgeting is useful for changes that will take place in a single year or in which the change is constant from year to year in a multiyear decision. However, partial budgeting should not be used for changes with uneven flows of costs and benefits that occur over several years. Producers should use investment analysis, which will be discussed later in this chapter, for such decisions.

Marginal Analysis and Response Curves

Fertilizer application is a good illustration of a basic economic concept of resource or input allocation. Most inputs follow the law of diminishing returns (figure 4-1), which states that at some point the

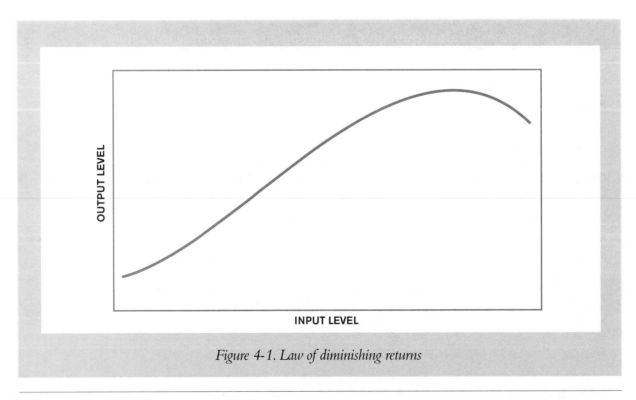

Figure 4-1. Law of diminishing returns

additional production from the use of an additional unit of an input will decline as more of the input is used.

In the fertilizer example, as nitrogen is applied to grass, grass yields will increase up to a point of maximum production. However, the additional yield from an extra pound of nitrogen will decline as nitrogen is added up to the point of maximum yield. If more nitrogen is added beyond that point, it is either wasted, having no effect on yield, or it actually hurts the grass plant and reduces yield.

Application of economics to this relationship is fairly simple. Combining the price per pound of nitrogen and a value per pound of grass with the pounds of nitrogen applied and the pounds of grass produced transforms the production relationship to one comparing the cost of applied nitrogen to the value of grass production. Under the law of diminishing returns, the value of grass produced will decline as nitrogen is added. Because the cost of nitrogen is the same for each additional pound, the economic optimal use of nitrogen would be that amount where the value of grass production from the last pound of nitrogen applied just equals the cost of that nitrogen. At points below that amount, a pound of nitrogen would return a greater value of the additional grass than its cost. At points above that amount, a pound of nitrogen would cost more

than the value of grass it produced. Producers should recognize that the amount of input use that maximizes profits will generally be less than the amount that maximizes production.

This type of analysis can be applied to other inputs as well. For example, it is the basis for the economic adjustment of livestock rations, where marginal feed costs are compared with the value of milk production or additional weight gain. A relevant case concerns supplemental feeding of grain to grazing dairy cows. For this analysis, the initial pound of grain will increase milk production 1 pound per day and will reduce pasture dry matter intake about 0.6 pound. If excess pasture production is already harvested, the reduced pasture intake will increase forage harvested by 0.6 pound. In addition, assume that the farm purchases hay for winter feeding. Further assume that feeding grain requires 2 minutes per day per cow or 0.2 minute per pound of grain for cows fed 10 pounds of grain and that feeding grain requires no further investment in equipment or operating costs for equipment.

First, let's look at $13.00 per hundredweight (cwt) for milk, $0.08 per pound for concentrate, $90 per ton for harvested forage, and $10 per hour for labor. The components of the budget for a pound of grain are shown in table 4-2.

Table 4-2: Grain Supplementation—1.0-Pound Milk Response			
BENEFITS		**COSTS/POUND OF GRAIN**	
Increased Revenue		*Decreased Revenue*	
Milk sales	$0.130	None	
Decreased Expenses		*Increased Expenses*	
Harvested forage	$0.027	Grain	$0.080
		Labor	$0.033
Total Benefits	$0.157	*Total Costs*	$0.113
Net Benefit	$0.044		

If the price of grain were $0.10 per pound, net returns would decrease to $0.024. With this higher grain price and a milk price of $9.00 per cwt, net returns would be –$0.016. Supplemental grain feeding is not always profitable; it depends on the price of milk and grain, as producers in other countries have discovered.

Increases in grain feeding may not be profitable even with the above base prices because of decreasing milk response. The next example illustrates the benefits of an additional pound of grain when grazing cows are being fed 10 pounds.

In this case, labor costs do not increase, but milk production increases only 0.8 pound. With milk selling for $13.00 per cwt (table 4–3), the pound of grain is more profitable, because there are no additional labor costs.

The final example (table 4-4) examines the costs and returns of feeding an additional pound of grain if the cow is being fed 20 pounds total. If the milk response for a pound of grain is 0.5 pound at $13.00 per cwt, the extra grain is still profitable, which indicates positive returns to a wide range of supplementation, given current milk and grain

Table 4-3: Grain Supplementation—0.8-Pound Milk Response

BENEFITS		COSTS/POUND OF GRAIN	
Increased Revenue		**Decreased Revenue**	
Milk sales	$0.104	None	
Decreased Expenses		**Increased Expenses**	
Harvested forage	$0.027	Grain	$0.08
		Labor	None
Total Benefits	$0.131	**Total Costs**	$0.08
Net Benefit	$0.051		

Table 4-4: Grain Supplementation—0.5-Pound Milk Response

BENEFITS		COSTS/POUND OF GRAIN	
Increased Revenue		**Decreased Revenue**	
Milk sales	$0.065	None	
Decreased Expenses		**Increased Expenses**	
Harvested forage	$0.027	Grain	$0.08
		Labor	None
Total Benefits	$0.092	**Total Costs**	$0.08
Net Benefit	$0.012		

prices. However, it is obvious that the returns will not be positive if the milk response is much less than 0.5 pound.

In practice on the farm, it is usually difficult to apply marginal analysis at the level of detail described in the previous examples. It is discussed here, however, so that farm managers are aware of the relationship and can examine their use of various inputs where appropriate, such as for feeding grain to dairy cows. In the fertilizer example, an actual optimal allocation of nitrogen on grass pasture may be about 100 pounds per acre. This optimal allocation may be unknown to the farm manager, but he or she may still notice good yield response when the nitrogen application rate is increased from 80 to 120 pounds per acre. With this response, the value of additional grass production is greater than the cost of the additional 40 pounds of nitrogen per acre. This analysis suggests that the additional nitrogen seems like an appropriate strategy, which it is. However, this strategy is inefficient in that 20 pounds of nitrogen are wasted, reducing profits from the optimal application by about $6 per acre if nitrogen costs $0.30 per pound. The excess fertilizer also may have some environmental costs. The important point is that a seemingly profitable level of input may mask some unnecessary, hidden expense.

Machinery Costs

Machinery costs can be included in partial budget analysis, but care should be taken to ensure their accuracy, especially when the fixed costs of machinery ownership are considered. Variable costs of machinery use—such as fuel, oil, most repairs, and perhaps hired labor—recur annually, are straightforward, and can be valued from your own farm records. Ownership costs require more thoughtful analysis. Changes in machinery ownership being considered in a decision have an impact on the farm business for several years. As

mentioned previously, partial budgeting is not the best tool for decisions that affect more than one or two years. In the hay field to pasture conversion example discussed previously, unneeded hay equipment may be sold and fencing purchased to convert the hay field to usable pasture. In that example, the costs of owning a hay baler over its remaining useful life would no longer be incurred after converting to grazing, benefiting the farm for several years. Similarly, the fencing investment must also be allocated over the fence's useful life. Including average annual values for asset ownership costs in the partial analysis will serve as a reasonable estimate of the single-year expense of machinery ownership.

Asset ownership costs have four primary components. The first is depreciation, which is the reduction in the asset value due to use during a year. The second is often called "interest" but is actually an opportunity cost of having money tied up in the asset. The third cost is insurance, and the fourth is shelter. Insurance expenses are easily identified from farm records, and shelter may or may not be applicable to a particular decision because of existing facilities.

Annual depreciation savings can be estimated using the formula,

$$(M - S)/n,$$

where M is the current market price of the asset and S is the estimated value of the asset after n more years of useful life. The formula calculates the average decline in market value over the remaining useful life of the asset. For example, if a hay baler was worth $12,000 and could be used for another six years, when it would be worth $3,000, then the annual depreciation would be $1,500 [i.e., (12,000 – 3,000)/6]. Note that the remaining useful life will be very short for machinery that is worn and will soon need to be salvaged. Also, this treatment of depreciation is an estimate of the average decrease in the value of machinery through annual use over n years. It is not the same as the depreciation that would be used for tax purposes.

The opportunity cost of the asset will also change over time as the remaining value of the asset declines as it depreciates. Therefore, an average investment value of the asset is used for calculation of the opportunity cost. The formula is

$$[(M + S)/2] \times i,$$

where i is the cost of funds used by the farm. The value used for i will vary among farms. The opportunity cost, or what the investment would earn in another investment, is what should be used. An appropriate interest rate would be the short-term borrowing rate available to the farm if the alternative use of funds is to repay debt or to deposit in CDs. If the money would be invested in a more profitable manner, the appropriate rate could be approximated by adding a small premium, such as 2–3%, to the borrowing rate. With current borrowing rates for farmers, 10% would be appropriate in the latter case. Using the baler in the depreciation example, the annual opportunity cost of owning the baler would be $750 with a 10% interest rate $\{[(12,000 + 3,000)/2] \times 0.10\}$.

An alternative method for calculating the cost savings from selling the baler would be to use local custom rates for baling. Note that these rates include the variable costs of operating the baler (fuel, labor, etc.) and an additional margin to cover the fixed costs of the machinery and some profit for the custom operator. The advantage of custom rates is that they are a market price for use of the machinery rather than values calculated with numerous assumptions. Like all costs, ownership costs for machinery and other assets should be included in the analysis of alternative production practices only if they will change. If the baler used in the example will remain on the farm, ownership costs will not change on the farm. However, if the baler were sold, then the farm would save $2,250 per year in those economic costs. Similarly, other assets and expenses that may

not be used after a change to grazing would include feeding equipment, feed storage facilities, and labor. These should be included as cost savings only if the charges are no longer incurred. For example, if feed storage facilities were still maintained on the farm, even though not used, there would be no cost reduction for their investment costs.

Labor Expenses

Labor and management are difficult expenses to include in farm economic analyses because they are relatively fixed on many small farms and are often unpaid. However, labor has value, even if unpaid. Changes in hired labor expenses are straightforward in farm analyses because hours and wage rates are known or can be estimated. Unpaid family and owner-operator labor can be valued or omitted from the economic analysis on each farm, depending on the farmer's preference. If included, unpaid labor should be valued at a wage rate equivalent to the opportunity cost of the labor, which is what the labor would produce in another use. Typically, a wage is used that could be earned off-farm or would be paid for hired labor used instead of family labor. Note that nonwage employee expenses should also be included when valuing labor, whether paid or unpaid. These expenses would include payroll taxes (such as Social Security) and fringe benefits (such as insurance and free farm produce). Regardless of whether labor expenses are included in farm analyses, farm operators should be aware of any changes in labor and management requirements associated with any proposed change on the farm. Changes in these requirements are particularly important if limited labor is available, or if nonwork leisure activities that would be affected by farm production changes (hunting, family recreation) are highly valued.

WHOLE-FARM DECISIONS

Partial budgeting may be inadequate for large-scale changes in farm operations. Such changes would include farm expansions, addition of new enterprises, significant farm investments, or any proposed alternative that has multiple components. Situations with multiple changes require care in order to ensure that all changes are profitable. For example, adoption of management-intensive grazing involves changes in all the enterprises on the farm, so it is useful to analyze the whole farm situation rather than only the changes. The whole set of changes often results in greater profitability on a farm. However, one or more of the changes may be losing money, a loss that may be hidden by the overall profitability of the other changes. Whole-farm budgeting can be used to analyze these multiple changes.

Whole-Farm Budgeting

Whole-farm budgeting is a projection of all farm revenues and expenses for a year. A good starting point is the Schedule F form included by farm operations with their federal tax returns. This schedule itemizes farm revenues and expenses, usually on a cash basis, and depreciation is included, resulting in a simple whole-farm budget. However, a complete whole-farm budget should use changes in inventory to adjust the cash farm income. When whole-farm budgeting is used for decision-making purposes, inventory changes can be ignored if all farm production is valued as revenue and all expenses associated with that production are included. An example of a whole-farm budget is presented in table 4-5 (page 32) for conversion of a confined dairy feeding system to an intensive grazing feeding system. This budget is adapted from a case study of such a conversion (6). Here, year one is the confinement system, and year two is the grazing system. The last line in table 4-5 (page 32) indicates that net revenues increased with grazing from $117,600 to $141,000. This increase in net revenue occurs despite a decrease

in milk, cattle, and grain sales. These decreases are common with such a conversion—milk production decreases, culling rates decrease, and grain production decreases (or is eliminated). The decrease in revenue of nearly $70,000 would be surprising to many farmers and other individuals in agriculture; usually changes in production methods in agriculture result in increased production and sales. The net revenues increase because expenses decrease more than revenues. Most categories of expenses decrease, and none increase. Crop expenses such as seed, fertilizer, chemicals, repairs, and fuel decrease because less ground, if any, is used to grow crops. Veterinary expenses and some expenses associated with sales also decrease. The large dramatic expense decrease is for feed, which decreases over $59,000. This decrease in itself is more than the decrease in sales. Thus, this budget has an unusual outcome: the benefit is decreased expenses and the cost is decreased revenue.

This example obviously required the use of a whole-farm budget. Table 4-5 shows that the amount for most categories changes. These many changes are outside the usual relevance of partial budgeting. The whole-farm budget was important here because the decrease in milk production and sales would likely be a concern for many dairy farmers. Without documenting the decreases in expenses, the farmer would likely be unwilling to adopt this production system.

One may be concerned about sources of data for such a budget. Farm records would provide data on production relationships and revenues and expenses for projecting the confinement system in year one. Year two would require assumptions about changes in production and inputs with grazing. Results for case studies and research on grazing would be helpful to develop this budget. Like any planning process, the budget will not be completely accurate but will provide a projection of revenues, expenses, and net revenue.

Table 4-5: Sample Whole-Farm Budget for Conversion of a Confined Dairy Feeding System to an Intensive Grazing Feeding System

	YEAR 2 GRAZING SYSTEM	YEAR 1 CONFINEMENT SYSTEM
Income		
Milk	$240,000	$280,000
Cattle Sales	15,000	31,000
Grain Sales	0	13,000
Miscellaneous Receipts*	11,000	11,000
Total Income	**$266,000**	**$335,000**
Expenses		
Labor	$8,700	$9,400
Feed	43,500	102,700
Rent (Vehicles, Machinery, Equipment)*	100	100
Vehicle Expense*	1,000	1,000
Fuel	3,000	4,200
Breeding*	1,900	1,900
Veterinary and Medicine	3,500	4,700
Advertising*	2,400	2,400
Cattle Marketing	500	800
Supplies*	11,000	11,000
Fertilizer	300	4,300
Chemicals	600	7,100
Seed	1,000	2,600
Repairs	4,500	13,700
Storage	1,100	3,000
Freight	13,200	14,000
Records and Tax Preparation*	4,700	4,700
Land Rent*	12,000	12,000
Business Insurance*	2,500	2,500
Utilities	3,600	4,300
Association Dues*	2,700	2,700
Interest	400	700
Miscellaneous	2,800	7,600
Total Expenses	**$125,000**	**$217,400**
Net Revenue	**$141,000**	**$117,600**

*No change in the associated expense.

ENTERPRISE ANALYSIS

Many farms have several enterprises or cost centers, each of which should occasionally be examined for its profitability. Enterprise analysis is used to examine the profitability of individual enterprises and is useful in identifying problem areas that are masked by whole-farm analysis. A modern dairy farm with an agronomic program can be divided into several different enterprises, including the milking herd, the replacement animals, each forage crop, and each grain crop. In addition, cost centers might be established to track equipment costs and purchased feed expenses. When the enterprise analysis is complete, resulting costs and returns can be evaluated for profitability or compared with alternatives. For example, a farm manager may decide that buying replacement animals is less expensive than raising them on the farm; that buying corn is less expensive than growing it; that custom hay harvest is less expensive than owning and operating hay equipment; or that growing and marketing crops is more profitable than feeding them to dairy cows.

One way to do enterprise analysis is to include columns for each enterprise you want to examine to the right of the whole-farm analysis in the whole-farm budget. Then part of each revenue and expense item is allocated to individual enterprises or cost centers. Unallocated portions of expenses may be put in a farm overhead cost center. Note that these overhead costs must be reallocated later, or the profits from each enterprise must be interpreted as contributions to covering farm overhead.

Units of production should also be included for each enterprise, and costs per unit should be calculated so that comparisons with market prices or custom rates can be made. For example, the cost of growing corn should be divided by total bushels of corn produced to get a cost per bushel, which would then be compared with the cost of purchasing corn.

Three considerations in enterprise analysis merit further discussion. The first is that cost centers must be transferred to the production enterprises on a use basis. For example, if all purchased feed is allocated from the whole-farm column to a cost center, the dairy enterprise must "buy" the feed from the cost center at market value. If more in-depth analysis is desired, storage costs and losses may also be added in that cost center, increasing the cost to the dairy enterprise. Without this intermediate cost-transfer step, the dairy enterprise will seem to be more profitable than it really is because it incurs no expense for purchased feed.

Similarly, each of the crop-production enterprises should "sell" feed that is consumed on the farm to the appropriate livestock enterprises. Otherwise, no sales of hay, for example, would indicate that hay is unprofitable when it may not be, and dairy would get "free" hay, which has a production, or purchase, cost. Pasture should be valued at a price for hay of similar quality. The difference between pasture and hay is in the quantity harvested and the method of harvest. Harvest costs should differ between hay and pasture.

Finally, farm managers must take care in analyzing enterprises and drawing conclusions about farm profitability. Many of the fixed expenses associated with capital and human investment (such as land, equipment, buildings, and unpaid labor) can be allocated to different enterprises. However, it should be recognized that these fixed expenses might not be reduced if an enterprise is deleted. For example, if the cost of growing corn is higher than the market price, it may seem more profitable to purchase corn. If part of the cost of growing corn includes tractor expenses that will still be incurred by the farm, even if corn is not grown, then the cost of growing corn is perhaps overestimated for comparison purposes. In this case, the cost compared to the market price should be the contribution of corn to covering overhead expenses.

Enterprise analysis can be useful in identifying weaknesses or strengths among the various activities found on a modern farm operation.

However, most farm decisions are made using marginal analysis or whole-farm budgeting. Enterprise analysis is most useful in determining how your farm compares with the profitability of other farms through comparative studies of profitability and cost of production published by various land grant universities. Enterprise budgets are also useful in projecting revenues and expenses of new enterprises or production systems for both partial budgeting and whole-farm budgeting.

INVESTMENT ANALYSIS

Any potential investment must be evaluated using two criteria: profitability and feasibility. Profitability concerns the generation of more revenues than costs over the life of the investment. Feasibility refers to the generation of more cash inflows than cash outflows in each year of the life of the investment, or the ability to cover a shortfall in any time period (year or even month) from other sources of cash. Profitability refers to the economics of a decision while feasibility refers only to cash flow, often influenced by financing arrangements. Investment analysis is complicated, because most investments have flows of uneven amounts of costs and returns over many years. Because of the time value of money, future flows of dollars are not worth as much as those today. (A dollar received today can earn interest and therefore is worth more than a dollar received in some future time period.) In the previous example of the hay baler (page 29), the uneven flows of depreciation and opportunity cost associated with owning the baler were averaged. This method is often sufficient for relatively small investments on the farm. However, a more complete assessment of a farm investment would use net present value analysis.

Net Present Value

Net present value analysis is based on differences in the value of a dollar at different points in time. A dollar today has a present value (PV) of $1.

However, a year from now its future value (FV) is $1 + $1r, where r is the interest rate at which a dollar could be invested. For example, if the interest rate is 10%, then FV is $1.10. Another way to write the formula for next year's future value is $1(1 + r). After two years, FV is $1(1 + r)(1 + r), and so on. The future value equation for any year, i, can then be written as:

$$FV_i = PV(1 + r)^i.$$

We can then work backwards from the future, calculating the present value of a dollar received sometime in the future by rearranging this equation to solve for PV:

$$PV = FV_i/(1 + r)^i.$$

PV is calculated by multiplying a future dollar by a discount factor, d, which is equal to $1/(1 + r)^i$. For our one-year example, $d = 1/(1 + 0.1)^1 = 0.9091$. A dollar received next year, then, is worth only $1 × 0.9091 = $0.91 today. Similarly, a dollar received in two years would be multiplied by $d = 1/(1 + 0.1)^2 = 0.8264$ and would only be worth $1 × 0.8264 = $0.83 today. These factors are easy to calculate with modern hand-held calculators, especially financial calculators, or with computer spreadsheets.

An example of investment analysis is presented in table 4-6 (page 35). Information on the initial investment in fencing and annual costs and returns is taken from the partial budget decision presented in table 4-1 (page 25). Annual benefits of $2,405 are the net benefits from table 4-1 of $1,351 plus the annualized fence establishment costs of $1,054. In table 4-1, the annualized fencing costs were included as a typical yearly expense. In investment analysis, the investment in the fencing is included in the year that it occurs rather than amortizing over the life of the fence. The net benefits appear in the column in table 4-6 labeled "Income." Each annual figure in the "Income" column is multiplied by the discount factor for that year. The discount factors are based on a 10% cost of funds allocated to this project. Because of the time value

Table 4-6: Net Present Value Analysis of the Pasture Decision in Table 4-1

YEAR	DISCOUNT FACTOR*	INCOME**	PRESENT VALUE
0	1.0000	–$4,215	–$4,215
1	0.9091	$2,405	$2,186
2	0.8264	$2,405	$1,988
3	0.7513	$2,405	$1,807
4	0.6830	$2,405	$1,643
5	0.6209	$2,405	$1,493
Net Present Value			**$4,902**

*The discount factor is based on a 10% discount rate.

**The initial investment is shown as a negative. Incomes in years 1–5 are the net benefit of pasture shown in table 4-1 (page 25) plus the annualized fence establishment costs, so that fencing costs are not counted twice.

of money, a dollar received in year five, for example, would be worth only 62.09 cents today, meaning that the $2,405 received in that year is worth only $1,493 today. Adding the column of present values results in a net present value of $4,902. This number is greater than zero, which means that the investment returns more than the initial investment and the cost of the funds used in it.

Profitability versus Feasibility

Net present value analysis measures the profitability of an investment. However, not all investments have positive cash flows in every year. In our fencing example, an initial investment of $4,215 is required to buy and erect the fencing. While the analysis shows that the investment is profitable, without available cash or loans the fencing cannot be purchased and the investment is infeasible. If the fence is financed with a loan, the project will be feasible if the loan payments are less than the annual cash returns. The loan payments are calculated to amortize the amount of the loan or investment. Recall that the annual cash flows were $1,351 while the amortized fence costs were $1,054 in our example. Thus, the fencing is feasible

if a loan can be obtained to purchase the fencing materials. Alternatively, the sale of the baler would likely finance all or part of the fencing investment.

Many issues in investment analysis merit further discussion. For example, choice of the discount rate, comparisons of investments with different economic lives, treatment of financing, and inflation all complicate net present value analysis. However, a discussion of these topics is beyond the scope of this chapter. Readers are encouraged to read farm management and financial management texts to learn about these more difficult issues. See the references for this chapter in the bibliography at the end of this volume.

WHOLE-FARM FINANCIAL ANALYSIS

Three types of measures are used to evaluate the financial health of a business: solvency, profitability, and liquidity. Solvency is simply whether the farm has more assets than debt. Does the farm have a positive net worth? Profitability measures whether the farm makes money in any defined time period. Is net farm income positive? Liquidity measures whether the farm has enough cash to meet its cash obligations. Is the farm's net cash flow positive? Note that it is not simply enough to know whether a farm is solvent, profitable, and liquid. Levels of these measurements are important, too. Again, a complete discussion of these concepts is beyond the scope of this chapter. Readers are directed to farm management texts and the Recommendations of the Farm Financial Standards Council for more information. Grazing will hopefully improve these measures and farm financial health.

Solvency

There are several measures of solvency. One common method is simply a direct calculation of net worth, where debts are subtracted from the value of the farm's assets. These numbers appear on a farm's balance sheet. However, a simple level of net worth is not a good measure of solvency of the farm business. For example, a net worth of $100,000 is pretty good for a farm with total assets of $150,000, but not so good for a farm with total assets over $1 million. Consequently, we calculate a ratio comparing debts and assets. One such ratio is the debt-to-asset ratio, which divides total debt by total asset value. In the previous example, the first farm's ratio would be $50,000/$150,000, or 0.33. The second farm's ratio would be $900,000/$1,000,000, or 0.90. Generally, a debt-to-asset ratio below 0.3 presents no worries; above 0.6 is cause for concern, and between these two levels should be watched carefully.

It is important to note a few additional points. First, farms at different stages of establishment will have different ratios. Young farmers will probably have more debt than older farmers who have paid off much of their debt. Therefore, what is sometimes more important than the ratio at any given time is the trend in that ratio over time. If the debt-to-asset ratio consistently increases over

time, the farm manager should be alerted to serious financial problems. Finally, the reason high debt-to-asset ratios are a cause for concern is that they measure the farm's risk exposure. One bad weather year or bad price year may cause the farm to lose so much money that total debt exceeds total assets. Monitoring solvency ratios gives farm managers insight into how exposed their farms are to risk.

Often, grazing operations can be managed with less investment in farm assets (machinery) so that less debt is required than similar-sized operations with confined-feeding programs and mechanically harvested crops. When converting from confined feeding to grazing, farm managers have the opportunity to improve their farm's solvency if assets are sold and debt is paid. In the partial budget example (page 25), farm solvency would not be improved unless the proceeds from the baler sale were used to pay off some debt. Otherwise, the proceeds would appear as cash in the asset column instead of machinery, and total assets would remain the same. Similarly, many livestock farms have feed storage structures (silos) that will not be used in the grazing operation. Asset values, total debt, or solvency does not change on these farms unless the silo is sold and sale proceeds are used to reduce debt.

Profitability

Profitability measures whether the business makes money in any time period, typically a calendar year. Net farm income is the most widely used measure of profitability. Note that all farm costs and returns must be included in the net farm income measure. It does not just measure cash in and cash out like a checkbook. Noncash costs such as depreciation are included, as are changes in the values of product and input inventories and levels of debt. These should also be calculated to adjust net cash income so that net farm income accurately reflects the costs and returns to the operation for that year.

Two other measures of profitability include the rate of return on assets and the rate of return on equity. These relate net farm income to the size of the business. The rate of return on assets is typically used to compare farms and efficient use of assets. It is calculated by adding the interest expense to and subtracting the value of family labor and management from net farm income, and dividing that amount by the average total assets for the year. The rate of return on equity simply divides net farm income minus the value of family labor by the average equity (net worth) level for the year. This measure is also used to accurately compare profitability among farms of any size and can be used to compare the rate of return generated on the owner's farm investment with other investment opportunities such as stocks and bonds.

Using the previous example in the solvency section, the smaller farm has total assets of $150,000, debt of $50,000, and equity of $100,000. Assume that the annual interest expense on the farm is $5,000, net farm income is $30,000, the value of family labor is $15,000, and asset and equity levels are roughly the same from the beginning of the year to the end. This farm's rate of return on assets would be [($30,000 + $5,000 − $15,000)/$150,000] × 100, or 13.3%. Its rate of return on equity would be ($15,000/$100,000) × 100, or 15%.

The larger farm in the example has total assets of $1 million, debt of $900,000, and equity of $100,000. Annual interest expense on the farm is $90,000, net farm income is $30,000, the value of family labor is $25,000, and asset or equity levels do not change over the year. This farm's rate of return on assets would be [($30,000 + $90,000 − $25,000)/$1,000,000] × 100, or 9.5%. Its rate of return on equity would be $5,000/$100,000, or 5%. These two farms have the same level of net farm income, but the larger farm uses its assets less efficiently and has a lower return on equity.

Grazing can have a twofold effect on farm profitability. If grazing is more profitable than the system it replaces on the farm, then net farm income and the rates of return on assets and equity will increase. Additionally, if solvency is improved through the sale of assets and there is an associated reduction in debt as grazing is adopted, interest expense will decrease and net farm income will increase, so the rates of return on equity will be amplified. Similarly, increased net farm income on a smaller asset base will improve the rate of return to assets.

Liquidity

Measures of liquidity should generally be calculated for the future, because after you find you don't have enough cash income to cover cash expenses it's usually too late to measure it. Therefore, a projected cash flow statement is probably the best tool to use to evaluate liquidity. Loan principal payments and withdrawals for family living expenses should also be included in the cash flow. Managers should probably build in a cushion of 10–20% of total cash obligations to account for unforeseen cash needs through the coming year.

Grazing can affect liquidity and cash flow in several ways. Often, a less intensive management style such as grazing can result in lower productivity and less cash revenue. For example, grazing dairy herds sometimes results in a reduction in milk production when farms change from confined systems to grazing systems. This reduction in cash revenue is usually more than offset by associated reductions in cash expenses on these farms. However, these changes must be projected during the planning process to ensure that the farm will generate enough cash to meet its needs.

RISK MANAGEMENT

To understand risk management, it is helpful to start with definitions. Risk has a slightly different meaning in economics and business than in popular usage. Dictionaries define risk as "possibility of loss or injury." Using this definition, risk in an agricultural context has to do with low yields, outputs, and output prices, or high input prices that cause farmers to earn lower profits or incur a loss rather than make a profit. Alternatively, the loss could refer to bankruptcy or other loss of the business. However, this interpretation has analysis problems, because it implies usual or normal yield output, prices, and profits, which do not exist in farming. Thus, the business view of risk is an event having a probability distribution of outcomes rather than a single certain outcome. Note that risk from this viewpoint could be a gain rather than a loss. *Existence of variability is the key here.* Considering risk rather than losses to be variable is actually more descriptive of risk management. Management to reduce risk usually reduces potential gains while limiting losses. Risk management is concerned with five major sources of risk: *production, marketing, financial, legal and environmental,* and *human resources.* These will be discussed in more detail later.

Risk Aversion

Risk aversion is a characteristic of human behavior in economic decisions. The basic concept is that humans prefer certain outcomes to risky outcomes and will pay or accept lower returns on average to avoid risk. Insurance is an example of paying to avoid risk, and keeping money in savings accounts rather than investing in common stocks is an example of accepting lower returns to avoid risk. The underlying motives for this behavior are complex and may be related to fear of losses or avoiding the expenditure of resources that adjusting plans requires.

People vary in their level of risk aversion. People who keep most of their savings in bank accounts are probably more risk averse than those who keep their savings in the stock market. All people in all situations are not risk averse. Interest in gambling is one example in which the opposite holds. In other cases, risk seems to be ignored. However, most people in a market situation seem to be risk averse.

Risk-Return Tradeoff

If most people are risk averse, then market opportunities usually reflect risk aversion. In these market settings, a risk-returns tradeoff exists. The tradeoff is that risk can be reduced only if returns are reduced. Examples of such tradeoffs are given above. Examples also exist in agricultural production—for example, production of oats. Oat yields have much less risk than corn yields, but the average profit from oats is also much lower. Another example is routine animal vaccination. This practice reduces the risk of disease and the associated loss in profits. However, profits in years that the disease would not occur are lower due to vaccination costs. Choices of practices that exhibit a risk-returns tradeoff involve individual levels of risk aversion—risk averse individuals usually produce oats.

Risk-returns tradeoffs do not exist for all decisions. In some cases, adoption of practices can improve returns and lower risk. Many agricultural innovations have this characteristic. Soil testing is an example in which using the test avoids situations in which yield would be lower due to nutrient deficiencies, and it also reduces the cost of fertilizer use when nutrients are adequate. Farmers who use soil testing often find that profits are also increased. The average increase in yield and reduction in fertilizer use in certain cases more than offsets the cost of the test and cost of more fertilizer.

Risk and Grazing

As mentioned previously, the five major sources of risk are production, marketing, financial, legal and environmental, and human resources. Most sources of risk are reduced with grazing. Production risk is the major source of concern when adopting grazing. Grazing dairy cows has unusual risk effects in that production risk can both decrease and increase in some situations.

The major source of potential decreased risk arises from variability in pasture production. Many farmers and others evaluating grazing for dairy production feel that farms that use pasture face greater weather risk than those using confined operations because of dependency on pasture availability for feed. In reality, modern grazing management also relies on supplemental feeds, both forage and concentrate, and good grazing managers will be able to adjust rations to maintain production if weather events reduce pasture productivity. Furthermore, if dry weather, for example, reduces pasture yields for a grazing operation and hay must be purchased, a confinement operation would also have reductions in hay and silage yields and would have to purchase forage. When cool-season grass is used for pasture, peak production is in the spring, while alfalfa and corn are more dependent on summer rainfall. Since dry weather is more likely to occur in the summer than in the spring, pasture is probably less subject to weather risk than harvested forages. In addition, the cow is a less expensive means of harvesting forage than machinery, so pastures or hay fields might be grazed that would be uneconomical to harvest mechanically. Given that risk is likely to be decreased with grazing, grazing may increase returns and decrease risk, unlike most risk management situations.

Potential increases in production risk can arise in seasonal dairy grazing. A common strategy used by grazers is to match seasonal forage production to animal needs. An example would be calving in the spring to match the feed needs of peak lactation to the excess feed availability of spring pasture growth. Spring calving systems may have two sources of risk. First, cows must have a 12-month calving interval to maintain this system, and calving late in the spring extends the breeding window for a 12-month calving interval into the heat of the summer. Breeding during the heat of summer decreases the chances of successfully impregnating cows in a timely manner. If this occurs, the seasonal system cannot be maintained without increased culling and replacement purchases. Second, a spring calving system will likely start the cows on pasture at the time of or near to peak lactation, which puts peak lactation at risk during the adjustment to a new feed source. A rule of thumb is that for every pound of milk lost from peak lactation, a 200-pound reduction in total milk for the lactation will occur. Like most sources of risk, both can be reduced with good management, which may have a cost. Thus, seasonal production with grazing will likely increase production risk.

The impact of grazing on marketing risk is limited. An effect on output price risk seems unlikely. However, some reduction occurs in reference to input prices. Because grazing operations tend to have fewer purchased inputs, perhaps including purchased feed, grazing operations will have less exposure to fluctuating market prices of purchased inputs than will confinement operations.

The impact of grazing on financial risk was partly described in a previous section. If a grazing farm has lower investment and also less debt, the farm will be more likely to meet debt service obligations and other fixed cash commitments with adverse production and prices than a confinement operation with more debt. Alternatively, the grazing operation may be able to carry a higher proportion of debt than the confined operation, but at the same level of financial risk. A related effect of less investment and lower fixed costs per unit of production is that profit margins per unit will be higher, which provides more margin to cover adverse market and weather events. However, if fewer units are produced on the farm (lower milk per cow), volume may be inadequate to cover debt service, family living, and other fixed expenses when gross margin per unit is reduced significantly. Any change in financial risk will depend on the change in cash available for debt service relative to fixed financial obligations.

The popular perception is that grazing reduces environmental risk because reducing cultivation reduces soil erosion and lowers fertilizer and pesticide applications. Although this is generally true, urine and feces from cattle on pasture can still have negative environmental effects, particularly around watering systems, shade, and other areas where cattle congregate (6). In addition, manure must still be spread from the confinement and milking facilities on farmland. The economics behind these issues rest on which crops to grow that the manure can be spread on, and the investment in manure handling equipment and/or storage structures.

Grazing will reduce the need for manure storage capacity, but managing the timing and allocation of spread manure on paddocks and other crops is important. If crops are eliminated for pastures, palatability of pastures can be affected greatly by manure, whether spread mechanically by the farmer or naturally by grazing livestock. Maintaining palatability in grazing systems may have some costs, which will vary by farm.

Nutrient management may become more important on grazing operations as herd size and stocking rates increase. As pasture usage becomes more intense, inorganic fertilizer may be more important in improving plant productivity, which eliminates part of the environmental advantage of pasture over crops. Therefore, as grazing becomes more intensive, producers will need to provide more management of environmental risk.

Human resource risk may also be slightly reduced by adoption of grazing for dairy cattle. Most importantly, machinery operations are reduced, which reduces the risk of injury from this source. In addition, risk is reduced because labor requirements are also likely reduced. If this reduction results in less hired labor, the risks associated with hired labor are reduced. If the operator labor is reduced, the farmer has more time for personal activities. This time could improve his or her health. In addition, the farmer may be able to spend more time with his or her spouse, which may reduce the risk of divorce. These points are only conceptual, and no research has been conducted on the impact of the adoption of grazing on human resource risk.

CONCLUSION

Grazing economics can be challenging to many managers who haven't given much thought to economics, but an economic approach to farm decision making is vital to the success of farm businesses. In summary, three simple principles should be remembered. First, additional returns should always be greater than the cost to achieve them. Second, all costs and returns must be considered for accurate decision making. Finally, managers should acknowledge how they approach risky decisions and be willing to accept lower returns in exchange for less variability in returns if they want to avoid risk. These simple rules will help farm managers improve their operations and achieve greater economic returns.

CHAPTER 5
Marketing Commercial Feeder Cattle

Phillip I. Osborne and James Y. Pritchard

The U.S. beef industry is made up of a series of segments that operate independently yet still depend upon each other. The production, processing, and consuming segments have all experienced tremendous structural change over the past decade due to mergers, integration, and marketing alliances. However, all the dollars passing through the various segments of the industry flow from the consumer. The future of the beef industry depends on its ability to supply consumers with the products they desire in form, quality, and value. Beef producers are responding to consumer demands by focusing more on production practices that target consistent quality in beef and beef products. As the industry progresses through the next century, producers will be faced with a number of opportunities and challenges. How they respond to the market signals will determine their competitive position in the marketplace.

The success of forage-livestock programs and their value will be measured against the ability of the producer to market livestock successfully and focus on consumer demands. Doing this will enhance profitability potential. Some will find success in holding costs down in a commodity market, while others may excel in a more intensive direct marketing of retail product to the consumer. The market will continue to be quality-driven, and small producers will have to be innovative in their approach to marketing forage and livestock.

In developing a marketing system, a major emphasis must be placed on operational efficiency and effective pricing. Operational efficiency includes the functions of assembling, processing, packaging, and distributing cattle. Pricing efficiency may be determined by how well the system reflects supply conditions and how accurately prices reflect buyer demand for quality and service.

Feeder-cattle production in the Northeast and Mid-Atlantic region is seasonal. Almost 78% of the feeder cattle in the region are marketed in March, April, September, and October. This concentration of marketing tends to overload the infrastructure for transportation, physical facilities, and buyer's capital requirements. The overloading and interaction of these forces often causes unwarranted price fluctuations.

GRADED FEEDER-CALF SALES

To provide some systematic approach to marketing feeder cattle, the southeastern states, and more recently the midwestern states, have adopted a number of marketing methods and procedures. The first comingled, graded feeder-calf sales were established to promote the cooperative marketing of feeder cattle produced in a local, two- or three-county area. One of the first demonstration feeder-cattle sales was conducted in Jackson's Mill, West Virginia, in 1932. Small farms, typical for beef herds in the region, comingled ownership of cattle as a means of offering for sale larger numbers and improved uniformity of the lots. The calves were sorted according to breed, sex, and weight and described with a house grade. The grades for describing the cattle varied by state or region, using descriptive terms such as prime, choice, select, or NC1, NC2, or NC3. Early grades used to sort and describe the cattle typically reflected the flesh condition of the cattle and the "eye appeal" of the pens being offered for sale. It was not until the early 1980s that the U.S. Department of Agriculture (USDA) established a uniform grading system for feeder cattle that tied frame and muscle to slaughter grades (addendum A, page 62). (See addendum B (page 65) for a glossary of feeder-cattle terms.)

The graded feeder-calf sales programs promoted improved genetics, feeding, and management practices, such as dehorning and castration, and used other marketing practices to attract buyers to regions and states where the graded sales were conducted. Well-organized feeder-calf sales continue to serve as an excellent means of providing small producers with a dependable market and providing feeder-cattle buyers with a reliable source of quality cattle.

The leadership and/or sponsorship of the graded sales has been a cooperative effort of the state cattlemen's associations, extension services, and state departments of agriculture. Area livestock associations, feeder-cattle marketing associations, or local stockyards often are responsible for the operation of the sales. Each state's graded feeder-cattle sales are required to operate under uniform rules and regulations approved by a state committee, but a market or association may establish its own rules and regulations that apply to a particular sale.

University or state Department of Agriculture employees have been responsible for the grading and sorting of feeder cattle. Graders and market reporters are trained and certified annually. The state veterinarian cooperates and oversees the interstate health regulations and activities to ensure the orderly transportation of feeder cattle.

State feeder-cattle marketing programs must be properly organized to be successful over a long period of time. They offer an excellent opportunity for the feeder-cattle producers, commodity organizations, and agricultural agencies to collaborate on the economic development of a region through improved marketing of feeder cattle. The organization of marketing associations often leads to other activities that promote rural development.

Recently, many seedstock producers have begun to develop market alliances and sponsor feeder-calf sales, primarily as a means of providing customer service to their bull-buying clients—particularly in areas where graded feeder-calf sales traditionally have not been held. State cattlemen's associations and the extension services have put together regional programs—such as the "Southeast Pride" or the Quality Assurance Feeder-Cattle Sales programs in Virginia, West Virginia, and Pennsylvania—that have special marketing aspects. Applying the same feeder sale concepts and packaging calves of similar genetics with additional source and process verification have allowed these programs to realize $42–83 or more per head for their cooperating clientele than would be realized from taking the same calves to the weekly auction markets. Some special feeder sales programs have not always experienced the same success. Sales volume, reputation, local demand, packaging, and market cycles still influence prices. The feeder-cattle sales programs have evolved into full-service marketing programs, providing participating producers with educational tools and opportunities, additional marketing alternatives, and a wealth of feedlot and carcass information.

TELEAUCTION: A VALUABLE MARKETING TOOL

The telephone has become quite a valuable marketing tool in special graded feeder-cattle sales. It has allowed order buyers and farmer feeders to participate actively with buyers in the ring and purchase cattle via conference call, thus avoiding costly travel and lodging expenses. Out-of-state buyers or even local buyers on tight schedules have appreciated the efficiency that the teleauction provides their business. It allows for efficient use of time and management skills, enabling them to do a larger volume of business in a shorter amount of time. Many farmer feeders like to use the system because they can continue to harvest crops and prepare to receive cattle at home. What makes this system feasible is a uniform grading system that provides the terminology to effectively describe the cattle and the volume of cattle that are available in the special graded sale. Comingling cattle allows producers to offer large uniform pens of cattle of similar weight, grade, breed, and sex, thus facilitating the ease of assembling trailer loads via the telephone. The teleauction has successfully been incorporated into the special graded barn sales, board sales, and video auctions in West Virginia and Virginia. The system is adaptable to all sales programs, because the telephone is familiar, dependable, and available to all.

Procedure for Conducting a Teleauction

The general procedure for conducting teleauctions is described below. See the sidebar for specific guidelines for feeder-cattle tel-o-auction field sales in Virginia.

Buyers who are interested in participating in a sale usually must notify the sale management two to

Guidelines for Virginia Feeder-Cattle Tel-O-Auction Field Sales

- Information on the cattle is available one week before the sale date from the Virginia Cattlemen's Association.

- Buyers must call the Virginia Cattlemen's Association 4 hours prior to the sale time to reserve a telephone hook-up and receive their bid number.

- Cattle are officially evaluated on USDA grade standards by a Virginia Department of Agriculture livestock marketing specialist, who also provides the estimated weight ranges, breed/colors, feed program, health program, flesh condition, cleanliness score, and weighing conditions.

- Buyers make payments for the cattle on the delivery date either by check or wire transfer of funds.

- An adjustment in price ("slide") is used if the actual weight of the cattle is heavier or lighter than the estimated weight of the load. The slide amount is announced at the beginning of each sale.

- The successful buyer, in cooperation with the seller, selects a delivery date. A buyer has up to 7 days following the sale to accept delivery of cattle unless otherwise stated.

- A buyer must give at least 24 hours' notice of pick-up time.

- Buyers are responsible for obtaining any and all necessary health papers to accompany cattle to their destination.

- The last bidder on "no sales" has the first purchase option at the owner's floor price.

three hours before the sale so proper reservations can be made. Interested buyers must provide proof of financial status and references to the livestock market prior to participating.

A listing or sale sheet describing each lot being offered is mailed or faxed to prospective buyers prior to the sale. A well-prepared sale sheet reduces the amount of time spent on the phone describing sale lots and greatly influences the efficiency of the sale.

A telephone conferencing service is used to facilitate the connection of multiple buyers to the auction. Each buyer calls the phone number assigned to the sale about five minutes before the scheduled auction time. Buyers are assigned a bid number, and a roll call is conducted to test the system and ensure that all interested parties are connected. A chairperson for the conference call conducts the teleauction as a third party between the auctioneer and the buyer. This person must understand the grading system and be able to communicate with the buyers accurately to describe and provide any background data about the cattle, such as health records, genetics, and performance data, when available. The chairperson serves as a ringman on the phone, relaying the bids to the auctioneer while tracking who has the contending bid on each lot of cattle.

During the auction, each lot of cattle is briefly described by the chairperson, noting any changes that are not on the original sale sheet. Buyers are able to participate actively in the sale, bidding with their assigned numbers on each lot as they are offered. At the conclusion of the sale, buyers are faxed their invoices, and information concerning transportation, health papers, and payment is arranged with the market.

GUIDELINES FOR SPECIAL STATE-GRADED SALES

All special state-graded feeder-calf sales operate with a set of management guidelines that ensure the quality of the feeder cattle. Many special sales have truly become "special" based on the qualifications necessary to participate. Some

Sample Guidelines for Special State-Graded Sales

1. All calves are farm-fresh (from farm of origin).

2. All calves are born between January 1 and May 1.

3. All male calves are castrated and healed.

4. All female calves are guaranteed open.

5. All calves are properly dehorned and healed.

6. There is no active pinkeye.

7. All calves are healthy and thrifty.

8. Health programs are often directed by specific sales.

9. All calves must fall within U.S. Department of Agriculture feeder-calf grades for #1 and #2 muscle.

10. All rejected calves are returned to the farm.

marketing associations and cattle pools have regulations that exceed state standards, and the cattle marketed generally perform above average in the feedlot or cooler. Some programs require specific immunization programs, 21- to 45-day weaning, source-verified genetics, and stringent control concerning the use of antibiotics and growth promotants. These programs develop a reputation for performance cattle that is targeted toward specific marketing demands. The qualifications are meant to protect the reputation of the program and provide guidance and direction for the participating producers. See the sidebar for sample guidelines that are commonly used by several regional sale programs in the southeastern and Mid-Atlantic states. Tables 5–1 and 5–2 (page 45) detail the steps involved in conducting special graded sales.

Table 5-1: Schedule of Events for Special Board or Teleauction Feeder-Cattle Sale as Conducted in West Virginia

Producer notifies market of interest in participating.

West Virginia Department of Agriculture grades cattle on the farm two weeks prior to sale.

Livestock market/extension specialist assembles data, develops sale sheets, and distributes to buyers via e-mail, fax, or mail.

Teleauction sale is conducted at the livestock market with extension specialist/agent assisting the auctioneer on the phone as conference chair.

Cattle are assembled from the farms and delivered to the scales, sorted, penned, and loaded on the buyer's truck by the livestock market. The livestock market is responsible for collection of payment from the buyer and disbursement to the producer.

Summary of sales is developed by West Virginia extension specialist/agent and distributed to producers and buyers.

Table 5-2: Events of a Special Graded Sale

Producers deliver calves to market.

Calves are graded, sorted, and penned according to sex, weight, grade, and breed.

Sale sheet is assembled, typed, and distributed to buyers. Calves are sold at public auction. Teleauction service is provided.

Livestock market facilitates delivery on buyers' trucks and collects payment and disbursement to consignors.

Extension specialist/agent summarizes sales and distributes information to producers and buyers.

MARKETING POOLS

Innovative livestock producers have adapted to the changes that have taken place in the industry and are beginning to take charge of developing production and marketing programs that address the issues and demands of the marketplace. A majority of U.S. beef producers has fewer than 30 brood cows, and the economies of scale improve dramatically when producers can pool their resources and marketing efforts. Cattle producers across the Southeast have adopted the concepts of the special sales and use teleauctions as a way to develop feeder-cattle marketing pools.

Sample Qualification for a West Virginia Marketing Pool

1. All calves must be sired by performance-tested bulls or have expected progeny differences (EPD) values above breed average for growth.

2. All calves must be born between January 1 and April 15.

3. All producers must be WVBQA-certified.

4. All calves and cows must be source-verified.

5. All calves must follow the WVBQA Gold Program (page 47).

6. All male calves must be guaranteed to be steers.

7. All heifers must be guaranteed to be open.

8. All calves must be 50% Angus.

9. All calves must be represented and marketed through a bonded livestock market.

Beef producers first began prevaccinating calves to better prepare them for the stress of weaning and transport to the feedyards. The calves were graded on the farm and assembled in the same fashion as for the special graded sales. Producers were prompted to pool calves because of the mechanics of delivering and receiving calves at the markets on sale day. Challenges in delivering calves to the auction barns included the long wait in line at the auction markets, long travel distances to markets, and the producers' inability to transport and deliver their entire calf crop to the sale barn. Producers wanted a system that would maintain the unique identity of their calves in the marketplace distinct from stockyard cattle not properly prepared to handle the ensuing stress.

Pooling allows small beef producers to become more competitive and to carve out a unique niche in the marketplace. Producers are moving from commodity marketing toward a value-based program with premiums and discounts. Procurements, shared ownership, and marketing programs are all functions of a pooling program. The future of the program is tied to increasing the focus on customers and cattle as individuals.

Marketing pools can function in a number of arenas. In the Mid-Atlantic region, "pooling" has allowed a producer to market feeder cattle, preconditioned calves, and source- and process-verified calves, or to participate in numerous retained-ownership programs. The seedstock industry has begun to embrace the concept as a means of supporting and providing service to their commercial customers. Producers who have adopted the concepts and participated in marketing pools have benefited economically and socially. The industry has benefited through improved efficiency due to advances in animal health and genetics.

See table 5-3 (page 48) for a comparison of West Virginia Beef Quality Assurance (WVBQA) pool sales versus graded sales. See the sidebar for a summary of the minimum requirements for participation in a West Virginia Quality Assurance Sale. WVBQA feeder-cattle sales offer high-quality, genetically source- and process-verified, quality assurance feeder cattle from 14 marketing pools in West Virginia and Pennsylvania. The program markets 7,000 feeder calves each year as the result of small producers working together to produce, market, and assemble load lots of calves. Tables 5-4 and 5-5 (page 49) show examples of WVBQA calf and yearling sale offerings, respectively.

West Virginia Quality Assurance Feeder-Calf Health Programs

West Virginia beef producers have an opportunity to participate in a statewide marketing program to identify and enhance the value of superior genetics and health management. The following programs are the minimum requirements to participate in the quality assurance sale.

West Virginia Gold Program — Vaccinated and Weaned

I. Vaccination Protocol
  A. IBR PI3 BVD BRSV Lepto 5
  B. 7-way Clostridial
  C. *H. somnus*
  D. **Pasteurella**

II. Additional Treatments
  E. Dewormed
  F. Grub and lice control
  G. **Booster** vaccinations for respiratory complex

Weaned for at least 40 days

Source-Verified
All calves within a pool are uniquely ear tagged, traceable to owners. The pools are interested in collecting performance and carcass data and are willing to cooperate with buyers to obtain data.

Performance Advantage Certification
Performance-tested sires from a central test or AI service must sire all calves, or calves must be sired by bulls that meet or exceed the breed average for weaning and yearling weight EPD. EPD minimum requirements for performance advantage certification for bulls born in 1997 and 1998:

Breed	Yearling Weight EPD	Weaning Weight EPD
Angus	56	31
Charolais	14	8
Gelbvieh	53	30
Hereford	52	31
Limousin	14	8

West Virginia Silver Program — Prevaccinated

I. Vaccination Protocol
  A. IBR PI3 BVD BRSV Lepto 5
  B. 7-way Clostridial
  C. *H. somnus*
  D. Pasteurella

II. Additional Treatments
  E. Dewormed
  F. Grub and lice control

Table 5-3: Comparison of WVBQA Pools and Graded Sales

West Virginia 2001*

2001 HEIFERS Weight Class	# Head	Avg. Weight (lb)	BQA Calf Pool $/CWT	BQA Calf Pool $/Head	Graded Barn Sale $/CWT	Graded Barn Sale $/Head	BQA Advantage ($)
300–399	22	375	93.28	350.01	85.49	320.78	29.23
400–499	299	464	91.78	426.21	84.62	392.97	33.24
500–599	783	545	90.08	491.27	79.35	432.76	58.51
600–699	514	645	87.01	561.12	75.59	487.50	73.62
700–799	117	739	83.46	616.67	70.87	523.65	93.02
Summary	1,735	572	88.74	507.42	78.14	446.83	60.59
Total Dollars			$880,370.97		$775,252.87		$105,118.10

2001 STEERS Weight Class	# Head	Avg. Weight (lb)	BQA Calf Pool $/CWT	BQA Calf Pool $/Head	Graded Barn Sale $/CWT	Graded Barn Sale $/Head	BQA Advantage ($)
300–399	19	373	105.75	394.61	109.53	408.72	(14.11)
400–499	244	465	105.28	489.90	100.41	467.24	22.67
500–599	974	554	99.71	552.82	91.25	505.93	46.90
600–699	1,084	645	95.41	615.01	83.21	536.38	78.63
700–799	435	738	90.44	667.27	79.19	584.27	82.99
800–899	123	839	80.31	673.55	76.53	641.89	31.66
900–999	23	911	78.70	717.03	71.45	650.97	66.05
Summary	2,902	622	95.42	593.30	85.57	532.07	61.23
Total Dollars			$1,721,751.54		$1,544,061.16		$177,690.38

2001 STEERS and HEIFERS

	# Head	Avg. Weight (lb)	BQA Calf Pool $/CWT	BQA Calf Pool $/Head	Graded Barn Sale $/CWT	Graded Barn Sale $/Head	BQA Advantage ($)
Summary	4,637	603	93.05	561.17	82.93	500.18	60.99
Total Dollars			$2,602,122.52		$2,319,314.03		$282,808.48

Values may vary due to rounding.

The difference between quality assurance sales and special graded sales is the guaranteed production practices and the reduced stress the calves experience in moving through the market channels. Incorporation of the board-sale concept reduces the amount of time cattle spend in assembling and shipping, often by days. Cattle assembled in the sale barns for the special sales often arrive in one day and are sorted, weighed, and penned by sex-weight-grade-color. The calves are auctioned the next day and, if transportation is available, are shipped within 4–5 hours following the sale. Due to the concentration of feedlots in the West, a majority of the calves face another 8–12 hours on the truck before arrival at the feedlots.

Table 5-4: Example of a WVBQA Calf Sale Offering

LOT 2	365 Steers, Morgantown, WV Pool			
Grade*		M1	L1	LM2
Breed	Blk	230	48	2
	BWF	56	9	1
	Char/SMK	3	5	1
	Red X	4	6	

Weight Range			
	Lot 2:	26 head: 450–510	Average weight: 480
	Lot 2a:	95 head: 490–560	Average weight: 525
	Lot 2b:	86 head: 550–580	Average weight: 575
	Lot 2c:	82 head: 580–640	Average weight: 605
	Lot 2d:	76 head: 630–720	Average weight: 660

Health Program	Gold Program (Boehringer Ingelheim Products) Dewormed: Ultramectrin Weaned Sept 1–3
Weight Condition	Cheslock Scales Morgantown 5- to 50-mile haul All calves weighed individually and sorted
Pickup Time	Oct 5–12
Flesh Condition	5–6
Feed	Pasture and preconditioning pellets
Comments	Excellent set of performance-tested feeder cattle. This group has pooled these cattle for 15 years, and quality and performance improves every year. All calves sired by performance-tested bulls. All producers are WVBQA-certified, and calves are source- and process-verified. There is plenty of uniformity, growth, and proven performance in this set of calves. Perfect cattle for a quality grid.

See addendum A (page 62) for grades.

Table 5-5: Example of West Virginia Yearling Board Sale Offering

Lot 2	59 Steers, West Virginia		
Grade		M1	L1
Breed	CharX	45	6
	RedX	6	2
Weight Range	800–925 Average weight, 860		
Health Program	Implanted, IBR PI₃ BVD, 7 Way Black Leg, BRSV, *H. somnus*, Pasteurella, Deloused/Grubicide, Dewormed		
Weight Condition	5-mile haul to Garrett scales		
Pickup Time	Within two weeks		
Flesh Condition	5–6		
Feed	Grass		
Comments	A very heavy-muscled, thick-bodied set of cattle will be a top set of feeding cattle. 5 steers weighing 950. No pink-nosed cattle. All are crossbred.		

The value of the board-sale concept is that the calves are graded on the farm and assembled for delivery at the buyer's convenience at a time when transportation is available. The calves are generally assembled in one morning and arrive at the feedyards the same evening, avoiding the costs of stress and poor performance. The reduction in the need for treating sick animals reduces the use of antibiotics and simultaneously enhances food safety, an issue that concerns the consumer.

The real advantage of marketing pools is the return of performance data, shared knowledge, and leadership development. The value is in the network of people working together to improve animal performance, profits, and the safety of an agricultural commodity they all take pride and pleasure in

producing. The marketing pools have provided a platform for the older or more experienced producers to serve as mentors for the younger or less experienced members of the group. The Northeast beef industry will benefit in the future from the leadership that has developed around the marketing concept. See the sidebar for a list of questions to consider when determining whether producers are potential members of a marketing pool.

FACTORS INFLUENCING VALUE

A number of factors affect the price of feeder cattle (see sidebar, page 51). The most obvious are sex, weight, and grade. Factors that affect producers' reputation, such as cattle health, weighing conditions, animal performance in the feedlot, and carcass quality on the rail, have a dramatic effect on price. Numerous studies have shown that the factors affecting the price of feeder cattle include sex, weight, number of head in a lot, grade, condition, breed, and presence of horns or other physical anomalies.

West Virginia data from the early 1980s showed a consistent $11–15 per hundredweight (cwt) discount for heifer calves. The primary reason for heifer discounts is lower average daily gains, lighter carcass weights, and lower dressing percentages than steers; some of the discount was built into the price due to the uncertainty of the pregnancy status. The introduction of board sales and teleauctions improved the value of yearling heifers, and the discount diminished to only $5–6 per cwt behind steer prices. The board sale heifer provides additional information to the buyers, and many of the producers were able to guarantee the heifers to be "open." Yearling stocker heifers will have carcasses about 80 pounds heavier than those put directly on feed as calves. Producers soon realized the value-added potential of backgrounding and grazing heifer calves, and the value difference between steer and heifer calves declined to an average of $8–10 per cwt in the 1990s. The

---

Are Producers Ready to Pool? An Assessment of Issues and Challenges

Consider the following questions and issues to determine whether producers are potential members of a marketing pool.

1. What level of commitment are they willing to make?

2. Identify the "movers and shakers" to rally the troops.

3. Those who are interested need to have demands and returns laid out in "dollars and cents."

4. If there is a high degree of "no interest," then more emphasis must be placed on education.

Factors Affecting the Value of Feeder Cattle

1. Breed composition
2. Grade
3. Sex
4. Weight
5. Feed cost for finishing cattle
6. Value of finished beef
7. Location/delivery

Value-Added Factors that Often Affect Price

1. Prescribed health program
2. Genetics
3. Reputation/performance
4. Volume
5. Uniformity
6. Weight condition

discounts between steers and heifers can still be quite variable; occasionally other factors, such as available feeder-cattle supplies and transportation cost, account for price differences.

The grade will affect the price of feeder cattle. In West Virginia and Virginia, the demand over the past 20 years has been for the medium-framed, #1-muscled cattle (see table 5-6, page 52). Even when the continental breeds were most popular, the medium-framed crossbreeds would bring more dollars per cwt. One important note about the cattle in the 1980s was that the large-framed, #1-muscled cattle brought the same price per head as the medium-framed cattle, but the small-framed cattle carried heavy discounts of $5–11 per cwt. By the late 1990s, the medium-framed,

#1-muscled cattle still demanded top dollar. The mix of cattle grades has resulted in more medium-framed cattle in the marketplace. The trend continues to move toward the medium-framed, heavier muscled cattle.

The large-framed cattle tend to be more popular in the western markets. The eastern markets favor the medium-framed calves, since they are most compatible with the yearling backgrounding and stocker programs. The eastern market has developed primarily around medium- and small-framed British-type cattle finished on higher forage diets than those in the western plains.

The shift associated with grades and breed types is best explained by the packers' and feeders' concern about carcass quality. Some breeds of cattle, such as Brahman and Herefords, have realized severe discounts in some markets (see tables 5-7, 5-8, and 5-9, pages 53–55). In most markets where carcass quality is an important part of pricing, Brahman breeding should be kept under 25%. New technology is becoming available to measure and predict quality, and pricing grids will likely cause a few shifts in visual appraisal of feeder cattle. Power is in performance information. Individual producers and marketing pools are beginning to assemble data about feedlot and carcass performance that should help avoid unnecessary discounts. The marketplace will be willing to reward the producers who are able to supply information concerning carcass performance and feedlot efficiency.

Table 5-10 (page 56) shows a statistical summary of factors affecting the value of a feeder steer across a six-year period in West Virginia. This summary shows the highest value for M1 and L1 steers of blackwhite face (BWF) or angus breeding sold early in the season with a premium for larger lot sizes. It also shows that even though average per cwt price ranged from $69 to $86 per cwt the marginal gain value was only $47-63 per cwt.

Table 5-6: Grade Comparisons

West Virginia Special Graded Calf Sales Fall 2001

STEERS				PRICE RANGE			
Grade	# Head	%	Avg Wt (lb)	Avg $/Hd	Min $/CWT	Avg $/CWT	Max $/CWT
L1	3,368	25	633	513.78	50.00	81.20	116.00
LM1	1,393	11	599	512.78	60.00	85.63	115.00
M1	6,500	49	555	501.51	53.00	90.40	126.00
S1	518	4	501	391.02	50.00	78.02	108.00
M2	1,471	11	529	441.46	42.50	83.40	112.00
LM2	6	0*	505	440.59	73.50	87.25	91.00
Tot/Avg	13,256		574	$494.80	$42.50	$86.16	$126.00

HEIFERS							
L1	2,325	22	579	439.84	47.50	75.92	92.50
LM1	1,019	10	566	437.27	62.00	77.21	88.00
M1	4,769	46	513	410.50	50.00	80.00	96.00
S1	482	5	462	302.26	35.00	65.45	89.00
M2	1,835	18	490	364.39	40.00	74.43	90.00
LM2	1	0	435	348.00	80.00	80.00	80.00
Tot/Avg	10,431		527	$406.54	$35.00	$77.21	$96.00

BULLS							
L1	120	27	561	420.73	60.00	74.97	105.00
M1	231	53	481	394.35	58.00	82.00	113.00
M2	70	16	434	338.16	57.00	77.97	105.00
S1	17	4	449	331.96	50.00	73.96	90.00
Tot/Avg	438		494	$390.18	$50.00	$78.96	$113.00

Due to rounding.

Table 5-7: Breed Comparisons

West Virginia Special Graded Calf Sales Fall 2001

STEERS

Breed	# Head	%	Avg. Wt. (lb)	Avg. $/Hd.	Min. $/CWT	Avg. $/CWT	Max. $/CWT
AngusX	5,869	44	569	506.49	54.00	89.08	125.00
BWF	4,249	32	586	501.65	42.50	85.65	126.00
Char	1,444	11	569	468.15	58.00	82.30	109.00
Exotic	759	6	582	477.07	53.00	81.93	106.00
Hereford	261	2	574	439.41	50.00	76.52	91.00
Other	234	2	567	455.91	53.00	80.40	105.50
RedX	336	3	543	440.74	50.00	81.14	110.00
Roan	74	1	578	479.20	65.00	82.89	104.00
Shrn	5	0	552	413.24	65.00	74.86	91.00
SimX	25	0	507	395.50	57.50	77.98	92.00
Tot/Avg	**13,256**		**574**	**$494.80**	**$42.50**	**$86.16**	**$126.00**

HEIFERS

Breed	# Head	%	Avg. Wt. (lb)	Avg. $/Hd.	Min. $/CWT	Avg. $/CWT	Max. $/CWT
AngusX	4,570	44	527	415.04	46.50	78.70	96.00
BWF	3,373	32	524	405.97	35.00	77.52	92.50
Char	1,119	11	524	389.43	47.50	74.39	90.00
Exotic	636	6	548	410.41	40.00	74.96	89.50
Hereford	154	1	516	345.24	43.00	66.88	84.00
Other	208	2	569	425.27	52.50	74.76	87.00
RedX	308	3	491	363.13	50.00	73.89	88.00
Roan	59	1	525	388.22	53.00	73.97	88.00
Shrn	1	0	530	307.40	58.00	58.00	58.00
SimX	3	0	453	344.15	73.00	75.92	82.00
Tot/Avg	**10,431**		**527**	**$406.54**	**$35.00**	**$77.21**	**$96.00**

BULLS

Breed	# Head	%	Avg. Wt. (lb)	Avg. $/Hd.	Min. $/CWT	Avg. $/CWT	Max. $/CWT
AngusX	111	25	510	410.22	69.00	80.50	105.00
BWF	168	38	489	397.15	57.00	81.15	113.00
Char	66	15	461	360.95	62.50	78.36	97.50
Exotic	35	8	520	391.14	58.00	75.18	93.00
Hereford	12	3	510	338.76	57.50	66.48	80.00
Other	12	3	536	396.29	60.00	73.90	105.00
RedX	29	7	477	358.24	50.00	75.15	95.00
Roan	5	1	531	384.16	62.00	72.35	85.00
Tot/Avg	**438**		**494**	**$390.18**	**$50.00**	**$78.96**	**$113.00**

Table 5-8: Breed By Weight Breed Summary

West Virginia Special Graded Calf Sales Fall 2001

STEERS

Pen Wt. (lb)	# Head	Avg Lot	Avg Wt. (lb)	Avg $/Hd	Min $/CWT	PRICE RANGE Avg $/CWT	Max $/CWT
AngusX							
400–499	1,538	14.4	466	464.06	70.00	99.55	125.00
500–599	2,042	23.2	556	505.72	54.00	90.89	107.50
600–699	1,385	21.0	645	536.41	54.00	83.16	95.50
700–799	396	10.2	746	586.58	59.00	78.65	88.00
BWF							
400–499	852	7.3	450	430.49	50.00	95.76	113.00
500–599	1,414	10.7	547	479.76	50.00	87.76	105.00
600–699	1,120	10.7	643	530.68	42.50	82.51	96.50
700–799	346	7.2	747	595.70	60.00	79.75	88.00
CharX							
400–499	303	4.1	453	409.57	68.00	90.31	109.00
500–599	509	5.7	547	454.76	60.00	83.11	94.71
600–699	375	5.2	639	510.01	58.00	79.77	91.50
700–799	105	3.4	735	561.13	59.00	76.30	85.00
Exotic							
400–499	143	6.0	459	412.85	58.00	90.01	104.00
500–599	269	10.0	553	463.27	56.00	83.70	95.00
600–699	235	13.8	645	514.58	71.50	79.74	86.00
700–799	49	4.5	742	563.82	60.00	76.01	85.00
Hereford							
400–499	56	1.9	454	369.59	63.00	81.48	91.00
500–599	95	2.6	552	425.46	50.00	77.12	88.00
600–699	42	2.2	651	504.02	55.00	77.47	86.00
700–799	26	3.3	739	530.45	53.00	71.74	78.50
Other							
400–499	65	2.5	455	400.94	53.00	88.16	99.00
500–599	59	2.4	544	446.58	64.00	82.12	89.00
600–699	38	1.8	646	489.00	59.50	75.68	87.50
700–799	25	2.1	746	565.69	63.00	75.83	79.50
RedX							
400–499	102	2.3	463	403.01	70.00	87.07	105.00
500–599	87	2.2	557	446.99	65.00	80.23	95.00
600–699	61	2.1	656	519.90	60.00	79.29	88.50
SUMMARY		7.2	574	**$494.80**	**$42.50**	**$86.16**	**$126.00**

Table 5-9: Breed By Weight Breed Summary

WV Special Graded Calf Sales Fall 2001

HEIFERS

Pen wt (lb)	# Head	Avg Lot	Avg Wt (lb)	Avg $/Head	Min $/CWT	PRICE RANGE Avg $/CWT	Max $/CWT
AngusX							
400–499	1,823	16.3	462	381.77	56.00	82.57	93.00
500–599	1,542	20.6	549	430.57	50.00	78.40	86.50
600–699	657	10.4	639	476.97	50.00	74.65	83.50
700–799	182	4.4	744	524.13	46.50	70.46	81.00
BWF							
400–499	1,140	9.4	454	369.12	47.00	81.39	92.50
500–599	1,109	10.6	545	416.78	50.00	76.43	85.50
600–699	622	8.8	633	473.14	49.50	74.75	83.50
700–799	80	2.8	746	533.22	40.00	71.53	79.50
Char							
400–499	354	4.2	455	352.65	52.00	77.44	86.00
500–599	396	5.1	544	402.55	57.00	73.95	85.50
600–699	182	3.4	638	454.40	58.50	71.17	81.00
700–799	43	2.3	746	528.47	50.00	70.82	79.25
Exotic							
400–499	181	6.2	461	356.73	49.00	77.40	89.00
500–599	264	10.6	554	412.16	45.00	74.43	82.25
600–699	126	10.5	642	480.74	70.00	74.90	79.00
700–799	26	3.7	747	527.41	45.50	70.59	77.00
Hereford							
400–499	54	2.1	450	304.82	43.00	67.68	84.00
500–599	39	1.8	550	350.97	48.00	63.83	78.00
600–699	22	1.8	634	417.13	50.00	65.81	73.50
700–799	13	2.6	725	536.52	71.50	74.04	75.00
Other							
400–499	55	2.1	453	336.69	52.50	74.35	84.50
500–599	60	2.6	548	400.88	55.00	73.11	78.00
600–699	36	2.6	639	451.37	60.00	70.62	78.50
700–799	36	6.0	741	598.08	62.00	80.73	83.25
RedX							
400–499	117	2.5	451	346.36	52.00	76.87	88.00
500–599	83	2.1	548	396.51	50.00	72.39	85.75
600–699	35	1.9	637	436.45	53.00	68.52	79.50
700–799	7	1.4	757	508.09	55.00	67.11	83.25
SUMMARY	**10,431**	**6.8**	**527**	**$406.54**	**$35.00**	**$77.21**	**$96.00**

Table 5-10: Average, Standard Deviation, and Range of Component Prices over Six-Year Period 1989–1994 for Feeder Cattle Sold on Board Sales in West Virginia

Component		Avg	Std	Min	Max
Basic Calf ($/head)		77.20	27.16	29.92	105.42
Weight on Calf ($/cwt)		57.52	5.93	47.22	63.03
Grade	M1 ($/head)	49.29	7.17	45.54	63.87
	L1	44.51	6.85	38.99	56.56
	S1	5.58	7.35	0.00	18.29
	LM2	0.00	0.00	0.00	0.00
Breed	BWF	25.72	8.67	14.65	36.59
	ANG	19.27	5.94	11.10	26.21
	CHARX	17.35	6.36	9.95	23.84
	EXOX	11.60	7.60	0.00	21.78
	HEREX	0.00	0.00	0.00	0.00
	HERE	-5.80	6.52	-12.97	0.00
Date	per day after Oct 1 ($/head/day)	-0.52	0.26	-0.87	-0.25
	per day before Oct 1 ($/head/day)	0.52	0.26	0.25	0.87
No. Head in Lot	per head in lot ($/head/head in lot)	0.81	0.05	0.71	0.86

Annual Breakout of Component Prices over the Six-Year Period 1989–1994 for Feeder Cattle Sold on Board Sales in West Virginia

Component		1989	1990	1991	1992	1993	1994
Basic calf ($/head)		93.95	93.75	105.42	29.92	70.35	69.82
Weight on Calf ($/cwt)	56.70	63.03	55.07	61.29	61.81	47.22	
Grade	M1 ($/head)	45.54	46.25	46.47	63.87	46.25	47.37
	L1	39.31	46.25	46.47	56.56	38.99	39.48
	S1	9.50	5.69	0.00	18.29	0.00	0.00
	LM2	0.00	0.00	0.00	0.00	0.00	0.00
Breed	BWF	14.65	18.70	21.68	33.07	29.62	36.59
	ANG	11.10	16.40	15.28	24.85	26.21	21.78
	CHARX	9.95	11.31	13.82	23.42	23.84	21.78
	EXOX	0.00	7.87	9.23	14.15	16.54	21.78
	HEREX	0.00	0.00	0.00	0.00	0.00	0.00
	HERE	-12.97	-8.95	-12.90	0.00	0.00	0.00
Date	per day after Oct 1 ($/head/day)	-0.27	-0.50	-0.25	-0.42	-0.79	-0.87
	per day before Oct 1 ($/head/day)	0.27	0.50	0.25	0.42	0.79	0.87
No. Head in Lot	per head in lot ($/head/head in lot)	0.80	0.71	0.86	0.80	0.81	0.85

Average Feeder-Cattle Weight, Price Per Pound, and Value Per Head for Six-Year Period 1989–1994

Year	1989	1990	1991	1992	1993	1994
Avg Value ($/head)	454	472	471	441	462	387
Avg Weight (lbs)	562	552	576	552	553	557
Avg Price ($/lb)	0.81	0.86	0.82	0.80	0.84	0.69

Feeder-cattle health problems always influence prices. Discounts for sick cattle are largely due to the high risk posed by these calves. "Stale" cattle receive average discounts between $5 and $10 per cwt. The special graded feeder sales will sort out unhealthy, eye-scarred, or other physically impaired cattle from the graded cattle. Those cattle are sold after the graded cattle, with severe discounts. Some markets will not accept "out" cattle for the special sales, and they are returned to the farm. Producers shipping calves to the sale should be sure that they travel in a clean bedded truck or trailer to arrive in a clean fresh manner. Eye appeal will increase the bids and reduce discounts associated with mud and manure tags.

Feeder-cattle buyers prefer to buy cattle that are in average flesh, discounting the overly fleshy or thin cattle. Data from The Ohio State University, Kansas State University, West Virginia University, and Virginia Tech have shown that season of the year and flesh condition affect the price. In the spring when cattle are being purchased for stocker programs, the thinner fleshed large and medium #2-muscled cattle will sell with little discount to large and medium #1 cattle. The above-average flesh cattle returning to grass have a tendency to gain less for the season. Yet in the fall, the fleshier cattle are not as severely discounted.

The stress of weaning, shipping, and weather requires that the calves have good body condition to draw on energy resources. When preconditioning or creep feeding calves, producers should be cautious to prevent overconditioning. Overconditioned calves tend to have more illnesses than average-fleshed calves. Some of this effect is likely due to changing the diet and the stress of movement to a different environment and feed program. Weaning programs incorporating at least half the diet as grazed forage have allowed better transition into backgrounding programs or direct movement to the feedyards.

Calves' weight influences their price. The differences in feeder-cattle prices are affected by the profitability of the feedlots and the stocker operations. The future of expected fed-cattle prices, corn prices, interest rates, and industry attitude also influence the price-weight relationship of feeder cattle. When feed is relatively cheap, the demand for lightweight feeder cattle increases, and more calves are placed directly into the feedlots. In a market where corn is costly, the demand for yearling cattle increases and favors the value-added situation of stocked cattle. In the early 1990s when corn was over $3.50 per bushel, a 900-pound yearling steer would bring the same per hundred as a 600-pound steer calf.

SELLING ON THE SLIDE

Using a "slide" when selling cattle adds flexibility to marketing. Cattle marketed on teleauctions/board sales, delivered on forward contracts, or sold using estimated weights are candidates to be marketed with a slide to protect both the buyer and seller. When several producers pool calves with different average weights that are sold at a single bid price, the slide can be used as a tool to treat each consignor fairly. An understanding of price spreads between various weights and classes of cattle is necessary to establish an equitable slide.

Slides can be one of the more difficult topics for many producers to understand when they sell cattle for later delivery. When cattle are sold and delivered at a later date, some provision must be made for minor adjustments in the price at delivery, should the average weight differ from what was advertised. Cattle may fail to hit the estimated weight at delivery due to a number of factors, including weather, unusual nutritional value of feedstuffs, unexpected performance, or human error in estimating weights. For these reasons, producers should strive to make the most accurate estimate of the delivery weight possible. This estimate should be based on past experience, test weight on all or a random sample of the cattle, and consultation with graders and livestock market personnel.

Yet even the best estimates sometimes miss the mark, and this is where the slide helps to resolve controversy. A slide is simply a formula to make adjustments in price based on corresponding changes in weight. The methodology of the slide must be determined prior to sale and announced at the time the cattle are sold. Remember, a slide is intended to make minor adjustments in sale price. *Do not assume that a slide will fix a problem where the weight estimates are significantly in error. A major error may require renegotiating the sale price.*

Five basic components are necessary to apply a slide:

Base weight	the expected average weight of the cattle at delivery time
Base price	the price paid by the buyer on sale day, based on the estimated weight
Window	the error allowed before any price adjustments are made (This is optional, and when it is used it is not more than 10–15 pounds.)
Actual weight	the actual average weight of the cattle at delivery time
Slide value	the per-unit adjustment that will be added to or subtracted from the price; the slide value must be sensitive to current market conditions

This is where confusion often begins. When solving a slide, it is easiest to calculate if you keep the price of the cattle in dollars per hundred pounds, such as $85.00 per cwt, and keep the slide adjustment in cents per pound, such as 5 cents per pound.

For example, a group of producers pools their calves, offers 75 head that they estimate will average 640 pounds when delivered, and receives $82.00 per cwt for their cattle at the sale. On the delivery date, the 75 head weigh a total of 48,852 pounds, an average of 651 pounds each. This weight is heavier than the advertised weight, and the buyer should pay a slightly smaller price. Using the slide we find that the cattle are 11 pounds heavier than the advertised weight (651 pounds − 640 pounds). Since we make the adjustment in cents per pound, 5 cents per pound times 11 pounds is 55 cents. We then subtract this adjustment from the bid price: $82.00 minus 55 cents equals $81.45 per cwt.

The same process works if the cattle are lighter than expected, except that we will add the adjustment, since lighter cattle should have a slightly higher unit value.

Base weight = 550 pounds
Base price = $94.00 per cwt
Actual weight = 535 pounds
Slide value = 5 cents per pound

The solution:
550 − 535 = 15 pounds
15 pounds x 5 cents per pound =
75 cents
$94.00 per cwt + 75 cents =
$94.75 per cwt

EVALUATING MARKETING ALTERNATIVES

Producers should evaluate their marketing alternatives every year to account for the cyclical nature of the market. The weather, Wall Street, or the press can create situations that will make retaining calves for a period of time more profitable and reduce losses to the operation. Every year offers different opportunities and challenges, so it is important to explore all available marketing options. The same potential to add value may exist when calves are in the cycle lows as when they are selling high. It all comes down to managing risk, given the seasonal trend or the current economic environment. A sound marketing strategy, coupled with a detailed budget, increases the predictability of the expected outcome. Consult with your tax advisor to weigh the effect of marketing two calf crops in one year should you decide to retain ownership. Do not let this be too much of a deterrent, because there are various strategies to avoiding a heavy tax burden.

A sound market plan is a tool that helps you and your operation succeed. The plan has to be developed with information that accounts for current, intermediate, and long-term market activity. Enhancing profit potential requires that the marketing plan be flexible and targeted to meeting the goals of the operation.

Marginal Value of Calf Gain

Producers should know how to evaluate the marginal value of gain from a specific management practice. At times, the value of the gain in a preconditioning or backgrounding program may be negated by the market value of the additional weight. For example, take a look at the value of the gains in calves marketed in 1985 compared to 2000 (tables 5-11 and 5-12).

Table 5-11: 1985 West Virginia Feeder-Calf Prices (actual weighted averages)

Weight Range (lbs.)	Avg. Wt. (lbs.)	$/Head	$/CWT
400–499	461	297.85	64.56
500–599	549	343.04	62.46
600–699	640	384.05	59.78

Value of the Gain from 450 to 650 Pounds

550 lb. @ 62.46 =	$343.53	1st 100 pounds of gain worth $53.01/head
450 lb. @ 64.56 =	$290.52	
650 lb. @ 59.78 =	$388.57	2nd 100 pounds of gain worth $45.04/head
550 lb. @ 62.46 =	$343.53	
		200 pounds of gain worth $98.05/head **OR $98.05/200 pounds = $49.02/cwt**

Table 5-12: 2000 West Virginia Fall Feeder-Steer Prices

Weight Range (lbs.)	Avg. Wt. (lbs.)	$/Head	$/CWT
400–499	461	485.14	105.22
500–599	553	506.43	91.56
600–699	647	545.77	84.30

Value of the Gain from 450 to 650 Pounds

550 lb. @ 91.56 =	$503.58	1st 100 pounds of gain worth $30.09/head
450 lb. @ 105.22 =	$473.49	
650 lb. @ 84.30 =	$547.95	2nd 100 pounds of gain worth $44.37/head
550 lb. @ 91.56 =	$503.58	
		200 pounds of gain worth $74.46/head **OR $74.46/200 pounds = $37.23/cwt**

In 1985, the value of the first 100 pounds from 450 to 550 pounds was $53.01. This means that a producer could add value to the calves by preconditioning or backgrounding the calves as long as costs were held under $0.53 per pound. The same is true if the calves were held to 650 pounds if costs were held under $0.49 per pound and the price of 650-pound calves remained at $59.78 per cwt. If the cattle were lightweights being bought in the spring to turn out on grass, holding the cost under $0.49 could be achieved more easily than with stored feeds.

In 2000, the same situation existed even when cattle were selling at higher prices. The cost of calves in 2000 experienced a decline from $105.22 per cwt at 450 pounds to $91.56 per cwt at 550 pounds. The first 100 pounds was worth only $30.09 per head. The market was willing to pay more for the heavier cattle, as illustrated by the value of the gain from 550 to 650 pounds ($44.37 per head). In 2000, the potential for added value was good as long as costs were held under $0.37 per cwt. If a producer were to background lightweight calves, again the best system appears to favor a grazing program.

Doing this exercise each year allows producers to evaluate the value of the gain before cattle enter a specific practice. Cow-calf producers can determine the value of retaining ownership to heavier weights versus the cost of gain for holding calves on available foodstuffs. Too often the value of lighter cattle is overestimated, failing to account for the declining dollars per cwt that heavier cattle will bring. The value of the gain from 450 to 650 pounds was not $105.22 or $91.56 per cwt, but only $37.23 per cwt. There are situations in which the added gain is cheaper to purchase than it is to produce.

MARKETING NUANCES

Volume and lot size have always had an impact on feeder-cattle prices. Buyers have always paid more for the larger pens of uniform cattle in the special graded sales. The teleauctions and board sales have generally most successfully marketed trailer-load lots of 48,000–50,000 pounds. The quality assurance calf sales have experienced the same phenomenon of the heavier calves going to the feedlots. Buyers of lightweight calves (under 550 pounds) prefer the smaller lot sizes of 30–50 head; they are generally going to assemble numbers that allow them to sell a trailer-load of yearlings the next fall. The local buyers of lightweight feeder calves are not so concerned with freight cost and are not having to fill tractor-trailer loads.

In previous studies of West Virginia graded feeder sales, the value of calves in the lot increased about $0.80 per head for every head in the lot. Kansas State University researchers found similar results— lighter calves sold in lots of 40 head brought as much as $7.50 per cwt over calves sold in lots of 10 head. However, yearling cattle sold in trailer-load lots of 70 head netted $4.50 per cwt premiums versus cattle sold in lots of 11 head.

Some markets have moved toward sexed feeder sales—all steer or all heifer—to take advantage of lot size where sales volume is too low. In a 500- to 600-head sale, the lot size will double if all cattle are of the same sex as compared with a mixed sex sale. Some producers complain about having to sort cattle at home or having to haul to two sales, but it is to their advantage in total dollars returned.

Shrink is an important factor when marketing feeder cattle. A description of weighting conditions should always be provided when marketing yearling cattle. Cattle that are weighed on the farm without a stand-in dry lot or a pencil shrink are usually discounted $2.00–4.00 per cwt. Cattle that are known to have good weights, resulting in minimal shrink upon arrival, have traditionally had more active bidding when the lots are offered. Uniformity in size and grade improves the eye appeal of a pen or load lot of cattle, and thus generates more buyer activity. Feeder farmers prefer the more uniform lots of cattle, because they like to empty pens all at once or in two market drafts. The genetic combination of feeder cattle is more important to uniformity of market animals than the weight and grade of the feeder cattle. Load lots of Angus cross calves marketed with less than 50-pound allowance from the lightest to the heaviest calf still yield carcass variability of 200 pounds. The use of ultrasound, video imaging, and computer analysis to compile data on cattle received in the feedyards shows promise in sorting cattle into uniform feeder groups, and far surpasses visual appraisal. However, feeder cattle have to be packaged into groups that can be described or appraised by the buyers. The USDA feeder grades and muscling scores provide an efficient method of packaging.

The feeder-cattle industry is going to have to position itself to provide a source and process that yields a verified, genetically predictable animal that consistently produces a product that supplies the consumer with a safe and favorable eating experience. Producers adopting the best technologies, incorporating a balanced management system, and allowing collection of accurate information about their product will hold a competitive advantage. They will best be able to respond to the challenges of the market and to be profitable.

U.S. Standards for Grades of Feeder Cattle

U.S. Department of Agriculture AMS-586, Agricultural Marketing Service, October 2000

The October 1, 2000, revision of the U.S. Standards for Grades of Feeder Cattle reflects changes in the genetic composition, production, marketing, and management of beef cattle since the 1979 grade standards were implemented. The updated frame size and muscle thickness grades more accurately reflect the value of today's feeder cattle.

The standards describe the various types of feeder cattle being produced and are used as a basis for federal–state livestock market reporting and as a common trade language between buyers and sellers. They are a tool for penning cattle at sales where feeder cattle are officially graded and ownership is commingled. They provide guidelines for better planning of breeding, management, and marketing programs. The feeder-cattle grades continue to be based on evaluating differences in frame size and muscle thickness—two of the most important genetic factors affecting merit (value) in feeder cattle.

Frame size is related to the weight at which, under normal feeding and management practices, an animal will produce a carcass that will grade Choice. Large-framed animals require a longer time in the feedlot to reach a given grade and will weigh more than a small-framed animal would weigh at the same grade. Thickness is related to muscle-to-bone ratio and at a given degree of fatness to carcass yield grade. Thicker muscled animals will have more lean meat. The Feeder-Cattle Standards recognize three frame size grades and four muscle thickness grades.

In addition to twelve combinations (three frame sizes and four muscle thicknesses) of Feeder-Cattle Grades for thrifty animals, an Inferior grade exists for unthrifty animals. The Inferior grade includes

feeder cattle that are unthrifty because of mismanagement, disease, parasitism, or lack of feed. An animal that grades Inferior could qualify for a muscle thickness and frame size grade at a later date provided the unthrifty condition is corrected.

"Doubled-muscled" animals are included in the Inferior grade. Although such animals have a superior amount of muscle, they are graded U.S. Inferior because of their inability to produce carcasses with enough marbling to grade Choice.

FRAME

	Frame	Expected Weight (lbs.) to Grade Choice	
		Steers	Heifers
	Large +		
L	Large		
	Large –	1250	1150
	Medium +		
M	Medium		
	Medium –	1100	1000
	Small +	1100	1000
S	Small		
	Small –		

MUSCLE SCORE

Minimum Thickness	Degree of Thickness
1	Moderately thick
2	Tends to be slightly thick
3	Thin
4	

MUSCLE THICKNESS

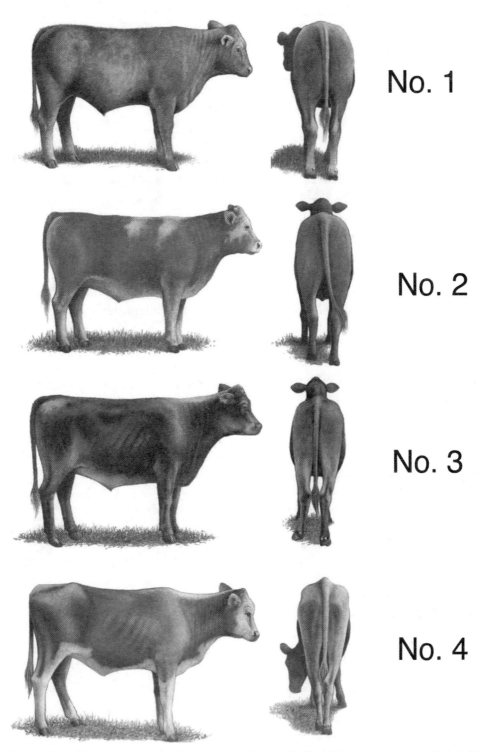

No. 1

No. 2

No. 3

No. 4

No. 1, No. 2, and No. 3 thickness pictures depict minimum grade requirements. The No. 4 picture represents an animal typical of the grade.

WHAT ARE THE FRAME SIZE STANDARDS?

Large Frame (L): Feeder cattle that possess typical minimum qualifications for this grade are thrifty, have large frames, and are tall and long-bodied for their age. Steers and heifers would not be expected to produce U.S. Choice carcasses (about 0.50 inch [1.3 cm] fat at twelfth rib) until their live weights exceed 1,250 pounds (567 kilograms) and 1,150 pounds (522 kilograms), respectively.

Medium Frame (M): Feeder cattle that possess typical minimum qualifications for this grade are thrifty, have slightly large frames, and are slightly tall and slightly long-bodied for their age. Steers and heifers would be expected to produce U.S. Choice carcasses (about 0.50 inch [1.3 centimeters] fat at twelfth rib) at live weights of 1,100–1,250 pounds (499–567 kilograms) and 1,000–1,150 pounds (454–522 kilograms), respectively.

Small Frame (S): Feeder cattle included in this grade are thrifty, have small frames, and are shorter bodied and not as tall as specified as the minimum for the Medium Frame grade. Steers and heifers would be expected to produce U.S. Choice carcasses (about 0.50 inch [1.3 centimeters] fat at twelfth rib) at live weights of less than 1,100 pounds (499 kilograms) and 1,000 kilograms (454 kilograms), respectively.

WHAT ARE THE MUSCLE THICKNESS STANDARDS?

No. 1: Feeder cattle that possess minimum qualifications for this grade usually display predominant beef breeding. They must be thrifty and moderately thick throughout. They are moderately thick and full in the forearm and gaskin, showing a rounded appearance through the back and loin with moderate width between the legs, both front and rear. Cattle show this thickness with a slightly thin covering of fat; however, cattle eligible for this grade may carry varying degrees of fat.

No. 2: Feeder cattle that possess minimum qualifications for this grade usually show a high proportion of beef breeding and slight dairy breeding may be detected. They must be thrifty and tend to be slightly thick throughout. They tend to be slightly thick and full in the forearm and gaskin, showing a rounded appearance through the back and loin with slight width between the legs, both front and rear. Cattle show this thickness with a slightly thin covering of fat; however, cattle eligible for this grade may carry varying degrees of fat.

No. 3: Feeder cattle that possess minimum qualifications for this grade are thrifty and thin through the forequarter and the middle part of the rounds. The forearm and gaskin are thin, and the back and loin have a sunken appearance. The legs are set close together, both front and rear. Cattle show this narrowness with a slightly thin covering of fat; however, cattle eligible for this grade may carry varying degrees of fat.

No. 4: Feeder cattle included in this grade are thrifty animals that have less thickness than the minimum requirements specified for the No. 3 grade.

Metric numbers are approximate due to rounding.

Addendum B to Chapter 5:
Glossary of Feeder Cattle Marketing Terms

Backgrounding — The practice of purchasing weaned or unweaned calves and placing them on a growing ration in preparation for placement into a feedlot for final finishing or for turnout onto pasture.

Barn sale — A sale in which feeder calves are brought into a livestock market or other sale venue where they are sold at auction to the highest bidder.

Board sale — A type of sale in which cattle are graded and assembled into trailer-load-sized sale lots in advance of the day of sale. Cattle are sold based on the grading information and estimated weight, along with a detailed description of how the cattle will be weighed, the mechanism for adjusting final price (see *slide*), and a time window when the successful buyer can take possession of the cattle. Cattle are not physically transported to the sale facility for the auction but remain on the farm. Following the sale, the buyer and seller finalize arrangements to take delivery of the cattle. The term derived from when the cattle being offered for sale were described on a blackboard then sold off the board.

BQA sale — A sale in which all producers have successfully completed Beef Quality Assurance training and certify that the cattle offered for sale have been handled according to BQA guidelines.

Commingled sale — A type of graded calf sale in which calves are assembled into uniform lots based on sex, breed, weight, and grade. Each sale lot likely contains calves from different producers.

Expected progeny differences — The average difference in performance between the offspring of two animals of the same breed when each is mated to animals of the same average genetic merit.

Feeder cattle (feeders) — Cattle that are being sold with the intention of being fed for harvest.

Can be either calves recently weaned from their dams or yearlings that have been weaned, backgrounded, and grazed over the following grazing season.

Flesh — Refers to the overall body condition of "fatness" of feeder cattle. Calves that are excessively fleshy, or fat, will likely gain less in the backgrounding, stocking, or finishing phases and are discounted by buyers. When evaluating feeder cattle, care must be taken not to confuse flesh with muscle and vice versa.

Frame — Refers to the skeletal size of feeder cattle and is an indication of logical weight when harvested at the end of a typical finishing program.

Fresh cattle — Feeder cattle that are fresh from the farm.

Graded sale — Similar to a barn sale, except that feeder cattle are officially graded according to USDA feeder-calf grades.

Grid — Refers to value-based marketing of fed cattle in which premiums and discounts for changes in yield and quality graded can be displayed in a table. The unit price of a carcass can be determined by finding the intersection of quality grade and yield grade in the table.

Internet sale — A sale conducted over the Internet. Internet sales may post lots available and allow buyers to place and update bids over a preset time period or may use streaming audio and video to allow buyers to participate in real time.

Muscling — Expression of muscle mass in cattle. USDA feeder-calf grading standards place the most heavily muscled cattle as #1 to the least muscled cattle as #3.

Pencil shrink — A calculation done to reduce the weight of a load of cattle by a constant factor when they have not been physically shrunk by an overnight fast or trucking. A 40% pencil shrink on a 40,000-pound load is calculated this way: [40,000 pounds × (1–0.04)] = 38,400 pounds.

Pooling — Procedure in which several beef producers work together to market their calves. The actual mechanics of a pool can vary greatly from one pool to another, provided that there is consistency within the pool, all members agree to the same terms and practices, and the pool fulfills all claims that it makes about its cattle. In West Virginia, pools are highly organized and have evolved to the point that all producers are BQA-certified and all calves are uniquely identified, source- and process-verified, and either prevaccinated or preconditioned. West Virginia pools also strive to improve uniformity of the calves by using identical nutrition and health programs within a pool and by establishing standards for bull purchasing.

Preconditioned calves — Calves that have received the initial vaccination and boosters, have been weaned, and are ready to place on feed with minimal health problems.

Prevaccinated calves — Calves that have received the initial round of vaccinations prior to being sold.

Process-verified — A step beyond source verification: cattle are uniquely identified and all processes (vaccinations, treatments, implants, feed, minerals, etc.) are fully documented. Calves that have received individual treatment different from the group are noted.

Satellite sale — A sale in which audio and video of the sale are transmitted via satellite television, allowing buyers to participate in real time. Buyers use a teleauction to bid on cattle during the sale.

Slide — A procedure for adjusting the actual price of cattle at delivery.

Source-verified — Cattle are positively and uniquely identified and ownership can be traced back to birth of the calf.

Stale cattle — Feeder cattle that have changed hands numerous time between leaving the farm and arriving at the feedlot. Stale cattle pose a risk of increased sickness and poorer performance.

Stocker cattle (stockers) — Feeder cattle that are placed on pasture prior to entering the feedlot.

Teleauction — The use of the telephone to allow buyers to participate in an auction without needing to travel to the sale location. The teleauction was the first method allowing buyers to participate in real time during a sale from remote locations. The use of a conference call allows all buyers to hear the auctioneer and to bid in real time. Often a facilitator in the sale barn chairs the teleauction, receiving bids from buyers and relaying them to the auctioneer. The facilitator also should be prepared to answer questions raised by the call participants. A good facilitator understands cattle marketing and current market conditions, is informed about the sale, and is capable of addressing questions raised by buyers and providing detailed descriptions of the cattle, if needed.

Trailer load — The amount of cattle needed to fully load a tractor trailer. Typically 47,000–50,000 pounds of calves are considered a trailer-load lot. This would require approximately 87 head of 550-pound calves or approximately 60 head of 800-pound yearlings.

Video sale — A sale in which video footage of the cattle being sold is provided to buyers. This may be via videotape or other media sent to the buyer prior to the sale or via satellite link during the sale.

CHAPTER 6
Dairy Marketing

Geoffrey A. Benson

Milk is the main product for most dairy farmers, and the money it generates is critical to the financial viability of a dairy farm. This chapter approaches milk marketing from two very different perspectives—selling milk as a commodity and selling value-added dairy products in specialty or niche markets. The purpose of this chapter is to provide a description of these two main markets so producers can make sound business decisions about how and where to sell their milk and in what form.

"Marketing" has a different meaning for the farmer producing milk for the commodity market and the person producing dairy products for a specialty or niche market. The main marketing decision facing a farmer producing for the commodity market is choosing among the locally available dairy farmer cooperatives or other local milk buyers. For someone producing value-added products for a specialty market, marketing involves identifying potential customers and their preferences for specific products, packaging, and services; identifying marketing channels; making pricing decisions; and promoting and advertising the product(s).

Those who choose to sell milk as a commodity need to understand how their prices are determined in order to make sound business plans and decisions, including making an informed choice of milk buyer. Farm numbers are declining, and average herd size is increasing in the face of generally low and variable profits and the advantages of a larger scale of operation. Financial success depends mostly upon one farmer's ability to create a farm business that is more profitable than others. Milk markets are national in scope, and therefore, a farm must be competitive from a national perspective. As a commodity, the value of milk is determined by its components and quality characteristics, but there are some regional price differences and some differences among milk buyers within a region. Farmers selling milk as a commodity are largely price takers but can choose among competing buyers for their milk.

There are opportunities in value-added production and marketing, but the key to niche marketing is to provide specialized dairy products and services to consumers at a profit. This is a much more complex business than producing milk for the commodity market. It requires knowledge and skills in many areas of milk production, processing, and marketing.

All producers can benefit from a broad understanding of the commodity market for milk and dairy products. It is important to dairy farmers who sell milk as a commodity because milk sales are the primary source of income. An understanding of what drives farm prices is one key to effectively managing the business side of the dairy farm, including cash flow management, price risk management, and investment decisions. It is important to those who add value to their milk by further processing and any related services because sound financial management requires pricing raw milk on a market-value basis so the profitability of the various profit centers can be accurately calculated. Also, the price of raw milk affects the wholesale and retail prices of the high volume, mainstream dairy products that are a reference point for pricing specialty products sold in niche markets.

MARKETING MILK AS A COMMODITY

The prices U.S. dairy farmers receive for their milk and wholesale and retail prices for dairy products are the result of a complex interplay of forces, including government policies and regulations, consumer preferences, production technology, and many other factors. The U.S. dairy industry has been buffeted by political, economic, and regulatory changes in recent years, including changes in consumer behavior, milk production technology, and federal dairy policies and programs. The 1996 farm bill, the Federal Agriculture Improvement and Reform Act, initiated significant changes in federal milk market orders. The 2002 farm bill, the Farm Security and Rural Investment Act, continued previous policies and added a major new counter-cyclical payment program to support dairy farmers' incomes. So how can farmers anticipate the impact of these and future policy changes and economic events? This section discusses the forces affecting the dairy industry and milk prices. The intention is to give producers a basis for making informed business decisions and for evaluating policy issues and options.

The world economy is becoming ever more integrated, so it may be helpful to begin with a global perspective. The United States is the largest producer of cow's milk in the world, though it ranks second to the combined production of the countries of the European Union. Over 90% of world production is consumed in the country where it is produced. The European Union is the world's leading exporter, but these exports are heavily subsidized and represent surpluses created by generous government programs. New Zealand, a low-cost producer, is the second largest exporter of dairy products.

U.S. domestic prices normally are significantly higher than world prices, which are artificially depressed by European Union subsidies, so U.S. exports are limited to high-value products or products exported with assistance from U.S. government subsidies. For many years, the U.S. dairy industry was protected from cheaper imports by quotas. Under the World Trade Organization trade agreement of 1994, these quotas were converted to tariff-rate quotas with high over-quota tariffs, which have a similar effect in limiting imports. U.S. imports represent approximately 6% of total U.S. production (table 6-1), and exports

Table 6-1: U.S. Milk Supplies and Utilization, 2001	
Item Supply	Amount * billion pounds
U.S. production	165.3
Beginning commercial stocks	6.8
Imports	5.7
Total Supply*	**177.8**
Utilization	
Farm use	1.3
Exports	2.2
Government purchases, net	0.2
Ending commercial stocks	7.0
U.S. commercial use	167.2
Total Utilization	**177.9**

* *Milk equivalent, milk-fat basis*

Sources: Dairy Market News, *Agricultural Marketing Service, U.S. Department of Agriculture, vol. 69, report 17;* Livestock, Dairy and Poultry Outlook, *Economic Research Service, U.S. Department of Agriculture, LDP-M-96.*

represent 1–2%. In comparison, world production in 2003 was approximately 1,050 billion pounds. A new round of trade negotiations, the Doha Round, began in 2001 but, based on initial progress and the history of previous rounds, a new trade agreement is not expected for several years. In light of these circumstances, U.S. imports and exports will remain relatively small and, therefore, this section will focus on the domestic situation and issues.

The U.S. Market for Dairy Products

The customer is the boss, rich or poor, educated or ignorant, foolish or faddish. Therefore, it is important to understand the current market for dairy products and trends in the U.S. dairy industry.

Commercial use of dairy products, referred to in official U.S. Department of Agriculture (USDA) publications as *commercial disappearance*, increased at an average rate of 2.0% annually over the last 21 years (table 6-2), just a fraction faster than the 1.2% annual increase in the population during the same period (table 6-3). However, this slow increase in overall consumption masks a significant change in the product mix. Total fluid milk sales have increased very slowly over the past 21 years, while whole milk consumption has dropped from 80% of total fluid sales to less than half, as consumers switched to low-fat and skim milk. Butter sales have grown slowly, in part because of health concerns and in part because butter prices were significantly higher than margarine prices during the 1980s and early 1990s. Ice cream sales have grown slowly also.

Table 6-2: Commercial Disappearance of Selected Dairy Products, United States, 1980–2001

Item	1980 (million pounds)	2001 (million pounds)	Average Annual Change %
Fluid milk	53,006	55,097	+0.2
Ice cream	830	955	+0.7
Butter	895	1,268	+2.0
Cheese, natural	3,984	8,639	+5.6
Total Disappearance, All Products*	**119,049****	**169,418****	**+2.0**

*Milk equivalent, milk-fat basis.   **Total disappearance includes dairy products not listed above.*

Sources: Dairy Yearbook, 1995, *Economic Research Service, U.S. Department of Agriculture;* Dairy Market News, *Agricultural Marketing Service, U.S. Department of Agriculture, Vol. 69, Report 17;* Livestock, Dairy and Poultry Outlook, *Economic Research Service, U.S. Department of Agriculture, LDP-M-96.*

Table 6-3: Economic Factors Affecting U.S. Dairy Product Demand, 1980–2001

Item	1980	2001	Average Annual Change %
Population (millions)	225.6	284.5	+1.2
Disposable income per person ($1996)	$16,045	$23,687	+2.3
Consumer Price Index (CPI) – all items (1982–1984 = 100)	82.4	177.1	+5.5
CPI – all food (1982–1984 = 100)	86.8	173.1	+4.7
CPI – dairy products (1982–1984 = 100)	90.9	167.1	+4.0

Sources: U.S. Census Bureau; U.S. Department of Commerce; Bureau of Labor Statistics, U.S. Department of Labor.

The only items with a rapid rate of sales growth are yogurt and cheese. However, yogurt accounts for less than 1% of total milk consumption, whereas cheese accounts for approximately 36%, on a raw milk–equivalent basis. Cheese sales have shown strong and fairly steady growth, doubling over the 21-year period. Mozzarella use has grown most rapidly and now rivals American cheese in total production. Pizza, cheeseburgers, and other foods prepared or eaten away from home explain most of the increase. When viewed from this perspective, the growth in total dairy product sales has a rather narrow base.

Several factors play a role in dairy product sales. Based on information compiled from 1980 through 2001 (table 6-3, page 69), these factors include:

- Population growth at 1.2% per year means more mouths to feed.

- Dairy product prices have tended to increase more slowly than the overall rate of inflation and the prices of other food items. This helps dairy product sales because prices are lower both in real terms and relative to competing products.

- New substitute products represent a continuing challenge to traditional dairy products.

- Consumer purchasing power has grown, as measured by an annual average increase of 2.3% in real (inflation-adjusted) disposable income. Higher incomes allow people to eat more expensive foods, including livestock products. Also, as incomes rise, an increasing share of the consumer's food dollar is spent on convenience foods and food eaten away from home. This trend has contributed to increased consumption of cheese and some other dairy products. However, the economy had its ups and downs, with high inflation and low growth in the 1980s, a record period of economic growth with low inflation during the 1990s, and an economic slowdown in the early 2000s.

A variety of other influences affect consumer decisions. These include health and dietary concerns (real or imagined), which lead consumers to switch from high-fat to lower fat products and from animal fats to vegetable oils. Changing demographics include more single-parent and two-wage-earner families and a corresponding reduction in traditional methods of food preparation and eating, changes in age distribution, and changes in the racial mix of the population. Producer-funded generic advertising and promotion has had a measurable impact on consumer behavior.

In summary, there has been slow growth in total demand for dairy products in spite of several favorable factors. This slow growth reflects the combined effects of slow growth in most dairy product sales and significant growth in only one major product category—cheese. Continued slow overall growth in dairy product consumption seems to be the best the industry can hope for.

In an era of global markets, it is useful to examine international competitiveness and trade issues affecting dairy imports and exports. Agriculture was largely omitted from discussions on trade liberalization from the end of World War II until the 1980s, when a number of different trade negotiations were started. The North American Free Trade Agreement (NAFTA), which went into effect on January 1, 1994, has not had a major impact on the U.S. dairy industry. Restrictions remained in place on trade in dairy products between Canada and the United States. Trade restrictions between Mexico and the United States are being phased out gradually over 15 years. U.S. dairy product exports to Mexico dropped during that country's financial crisis, but have recovered slowly, with a positive but small impact on U.S. farm prices.

Market access and export subsidies were two key areas of the Uruguay Round, which created the World Trade Organization (WTO) and took effect on July 1, 1995. Reductions in global trade barriers and trade-distorting policies were phased in over a six-year period. Specifically, WTO member countries were required to allow a minimum level of imports equal to 5% of their domestic consumption. U.S. dairy imports increased from about 2% to 5% during this phase-in period. Import quotas had been used to control dairy imports since 1947, but these quotas were converted to tariff-rate quotas. This means a specified volume of dairy product imports must be allowed into the United States at a low (or no) tariff, but additional imports are subject to very high tariffs. U.S. dairy products have increased access to markets in other countries, but domestic prices are higher than world prices, which affects competitiveness, and there are limits to the export subsidies the United States may provide to assist dairy export efforts.

It seems clear that these changes have hurt rather than helped U.S. dairy farmers, but the impact has been relatively small. Domestic factors will continue to dominate the U.S. dairy economy for several more years.

Turning to the milk supply side of the equation, the U.S. dairy industry is a dynamic one, with a long history of increased productivity. However, increases in milk production per cow at an average of 2.5% per year (table 6-4) have exceeded the rate of growth in commercial sales. The resulting pressure on milk prices has forced a reduction in cow numbers and farm numbers. The decline in the number of farms has been even greater than the decrease in cow numbers. Based on data from the USDA, dairy farm numbers fell by 38% from 1993 to 2001, while cow numbers fell 5%.

While total U.S. production has been increasing slowly, marked regional shifts in market share have occurred, with gains in the West and loss of market share in the rest of the country (table 6-5).

Table 6-4: U.S. Milk Production, 1980–2001

Item	1980	2001	Average Annual Change (%)
Total milk production (billion pounds)	128.4	165.3	+1.4
Milk per cow (pounds)	11,891	18,139	+2.5
Cows (thousands)	10,799	9,115	− 0.7

Sources: Dairy Yearbook, *1995, Economic Research Service, U.S. Department of Agriculture;*
Livestock, Dairy and Poultry Outlook, *Economic Research Service, U.S. Department of Agriculture, LDP-M-96.*

Table 6-5: Regional Shares of U.S. Milk Production, 1980–2001

Region	1980 % of U.S. total*	2001 % of U.S. total*	Region	1980 % of U.S. total*	2001 % of U.S. total*
Northeast	20	17	Southeast	4	3
Lake States	29	22	Delta	2	1
Corn Belt	12	9	Southern Plains	4	4
Northern Plains	4	3	Mountain	5	13
Appalachia	7	4	Pacific	12	25

Totals may not add to 100 because of rounding.

Sources: Dairy Yearbook, 1995; Milk Production, *National Agricultural Statistics Service,*
U.S. Department of Agriculture, Da1-1 (2-02).

Farm-level production costs and returns vary year by year because of changes in milk prices, feed and other input costs, and efficiency gains (table 6-6). In general, productivity gains have resulted in costs and returns that have not kept pace with inflation. However, in any given year, there are wide differences in average costs and returns among the various regions of the country (table 6-7). The definitions of the individual regions are not directly comparable for tables 6-5 and 6-7, but it is safe to say that differences in profitability have driven the regional shifts shown in table 6-5. It should also be noted that there is great variation in farm size, type, and financial performance among farms within each region.

Table 6-6. Milk Production Costs and Returns, United States, 1980 and 2001

Item	1980	2001
	($ per 100 pounds)	
Gross value of production		
Milk	12.95	15.36
Livestock and other	1.38	1.91
Total Value of Production	**14.33**	**17.27**
Operating expense and overhead	9.41	11.45
Value of Production Less Listed Expenses	**4.92**	**5.82**
Capital recovery and charges for unpaid family-owned inputs	3.28	7.05
Total economic costs	12.69	18.50
Residual returns to management and risk	1.64	−1.23

Note: Costs and returns may not be strictly comparable across years because of changes in methodology.
Sources: Milk Costs and Returns, September 1996; 2000 Agricultural Resource Management Study, Economic Research Service, U.S. Department of Agriculture.

Table 6-7. Milk Production Costs and Returns, by Region of the United States, 2001

Region*	Value of Production	Operating Cost	Net Cash	Total Economic Cost	Returns to Management and Risk
		($ per 100 pounds)			
Heartland	17.90	10.95	6.95	22.96	−5.06
Northern Crescent	17.83	9.07	8.76	20.58	−2.75
Prairie Gateway	16.85	10.48	6.37	13.88	2.97
Eastern Uplands	18.49	13.07	5.42	25.88	−7.39
Southern Seaboard	19.69	10.35	9.34	18.34	1.35
Fruitful Rim	15.87	8.91	6.96	13.94	1.93

* For a description of the regions, see ERS Farm Resource Regions, at HTTP://WWW.ERS.USDA.GOV/PUBLICATIONS/AIB760/
Sources: 2000 Agricultural Resource Management Study, Economic Research Service, U.S. Department of Agriculture.

The Role of Government in Milk Marketing

Government involvement in milk production, pricing, and marketing is extensive and includes public health, federal price and income support programs; milk market orders; trade policy; domestic feeding programs; and humanitarian aid. Because of the impact of these programs on dairy farm incomes, an understanding of government involvement and programs is an important component of sound business planning and decision making.

Public health regulations determine how milk is produced and affect the cost of production. Federal dairy policy and federal orders affect dairy farm incomes and the financial risks associated with dairy farming. Government regulations specify that only Grade-A milk can be used for the production of fluid (beverage) milk products, to ensure a safe and high-quality milk supply. Grade-A dairy facilities must be constructed according to government specifications and then routinely inspected to ensure that the facility is maintained and operated according to specified standards. Finally, milk samples are taken on a regular basis and must meet certain quality standards if that milk is to be sold as Grade-A milk.

Many years ago, most milk was used for manufacturing dairy products, and many farmers did not need to comply with Grade-A regulations. However, largely because of the financial incentives provided by federal orders, most producers converted to Grade A, and today there is very little manufacturing-grade milk. Even farmers who process their own milk may find it advantageous to be certified Grade A in order to have a market for any excess milk.

The federal dairy price support program has had a significant effect on milk prices across the nation. The federal government has a long history of intervention in agricultural markets to address concerns about low farm prices and incomes, and this program has been a major component of dairy policy. The program began as part of the Agricultural Marketing Act of 1949, to "assure an adequate supply of wholesome milk at reasonable prices." During the 1949–1988 period, the support program raised farm milk prices and helped to stabilize them. However, at times the cost of the program has reached politically unacceptable levels, and the future of the price support program is uncertain. An understanding of the program operation is needed to evaluate the impact of future changes in farm policy on milk prices and dairy farm incomes.

Under this program, a support price for milk is set according to rules established by Congress, usually in the farm bills that began in 1949 and have been passed every four to six years since. These rules have changed from time to time, but the basic operation of the program has not.

The support price is announced in U.S. dollars per 100 pounds of manufacturing-grade milk. Because raw milk is highly perishable, it is not feasible for the government to intervene directly in raw milk markets. Therefore, the farm-level support price is converted into equivalent government purchase prices for storable dairy products—cheese, butter, and nonfat dried milk (NFDM).

The Commodity Credit Corporation (CCC), an agency of USDA, offers to buy these products at the announced purchase prices and must buy everything it is offered. If more milk is produced than consumers are willing to buy at prevailing prices, dairy processors can always sell products to the CCC. The basic idea is that the CCC is always there as a buyer of last resort to provide a floor under wholesale product prices. The government purchase prices consider manufacturing costs and product yields so that processors should be able to pay farmers the approximate support price in a competitive market for raw milk.

USDA has discretion over the price established for the different components of raw milk—butterfat and nonfat solids—as long as the combination achieves the mandated support price. In recent years, USDA has made periodic adjustments to the purchase prices for butter and nonfat milk solids to reduce the volume and cost of government purchases and to stimulate commercial sales. For example, the government purchased large amounts of butter during the early 1990s, so USDA reduced the butter price and offset this with an increase in the price of NFDM. NFDM purchases were large in the late 1990s and early 2000s, and USDA lowered the NFDM price and raised the butter price. These adjustments are often referred to as a change in the "tilt."

The justification for a federal price support program is usually expressed as a desire to stabilize and raise farm milk prices. The price paid to farmers for milk used to produce cheese has long been used as a benchmark that reflects the overall balance between milk production and sales. The relationship between the support price and the cheese milk price (Class III) is shown in figure 6-1. There was rapid inflation in the mid- to late-1970s and, at that time, the support price was indexed to a parity formula that was heavily influenced by the rate of inflation. From the mid-1970s through the 1980s the two prices were similar, as they have been over most of the life of the support program.

Largely as a result of the support price levels in the late 1970s, milk production increased and caused a large milk surplus in the early 1980s. The high cost of government purchases at a time when efforts were being made to control a large federal deficit led to changes in federal dairy policy. A series of

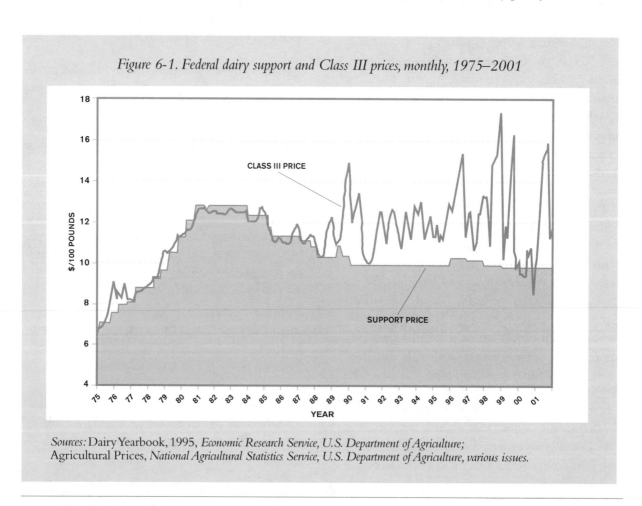

Figure 6-1. Federal dairy support and Class III prices, monthly, 1975–2001

Sources: Dairy Yearbook, 1995, *Economic Research Service, U.S. Department of Agriculture;* Agricultural Prices, *National Agricultural Statistics Service, U.S. Department of Agriculture, various issues.*

cuts was made in the support price to discourage production and reduce government outlays. Lower support prices translated into lower farm prices. In addition, a short-term milk diversion program was introduced that provided incentive payments to farmers who did not increase production. The Dairy Herd Buyout Program followed in 1986–1987, which paid producers to stop milk production for five years. A drought in 1988 and unfavorable weather in the northern dairy states in 1989 affected milk production. This combination of events gradually brought production and commercial sales into closer balance. By the end of the 1980s the support price had fallen below the average cost of producing milk, based on cost estimates published by USDA.

From 1989 through 2001, milk production and commercial sales were in close balance. Except for occasional, brief periods during some spring months when production is at a seasonal peak, market forces set prices rather than the price support program. CCC purchases were small compared with the levels seen in the 1980s. Market prices were above the support price, in general. However, as the national surplus decreased, milk prices became increasingly volatile, as shown in figure 6-1.

The farm milk prices that were generated by market forces during 1989–2001 were considerably higher than the federal support price, on average, but there were large fluctuations. Farm milk prices are very sensitive to small changes in production or sales, and these have a disproportionately large impact on farm prices. Milk production has a pronounced seasonal pattern that does not match the sales pattern (figure 6-2). These imbalances, combined with unpredictable variations in production and sales, created a price roller coaster with large month-to-month changes.

In times past, the price support program helped stabilize prices. Spring surpluses were removed, and if production fell and prices increased, dairy products could be resold from CCC stocks into the commercial markets, dampening price increases. The elimination of the dairy price support program would have little long-term

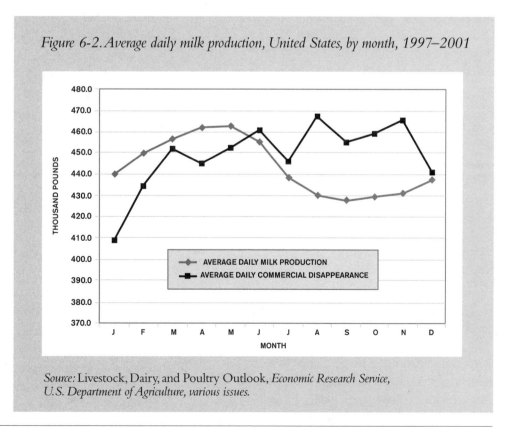

Figure 6-2. *Average daily milk production, United States, by month, 1997–2001*

Source: Livestock, Dairy, and Poultry Outlook, *Economic Research Service,* U.S. Department of Agriculture, various issues.

impact on farm prices, provided supply and demand remain in close balance. Ending the program would, however, remove a safety net that protects farmers from extremely low prices, which could occur if production increased unexpectedly or if sales fell, temporarily or permanently, and an increase in price volatility became likely.

Federal marketing orders were established under the Agricultural Marketing Agreement Act of 1937. The primary intent of this act was to provide a mechanism for solving problems of "disorderly marketing" in agricultural markets. This 1937 legislation built on previous legislation and established procedures for farmers to petition for, and vote on, the creation of marketing orders for agricultural products. It was believed that farmers were frequently placed at a disadvantage relative to the buyers of their products, because they lacked market power and because many farm products are perishable and must be sold quickly. Federal orders change the "terms of trade" in ways that are intended to assist producers. However, orders are not permitted to impose production controls.

The process of creating a federal order begins with a petition to the USDA. A public hearing is called and producers or their representatives are required to demonstrate that a marketing problem exists for a specified agricultural commodity in a certain geographic area and that a market order will help solve this problem. Any interested party — producers or their organizations, processors, and consumers — can testify at these hearings. The USDA then develops regulations based on the hearing testimony. The proposed regulations must be approved by a two-thirds majority of producers in a referendum. Once approved, a federal marketing order for a given agricultural product regulates the marketing of that product in a specified geographic area.

Most milk is marketed under federal milk marketing orders, and their rules affect the majority of the nation's dairy farmers. Federal milk

marketing orders (FMMOs) are administered by the USDA. The number of FMMOs and the area covered by individual orders has changed repeatedly over the years. Effective April 1, 2004, there were 10 FMMOs (figure 6-3, page 77), and these orders regulated approximately 60% of the Grade-A milk produced in the United States. California, the largest dairy-producing state, is the major exception, and this market is regulated under a state order that fulfills comparable functions. Well over 90% of the milk produced in the United States is Grade-A milk. However, in 2001, only one-third of Grade-A milk was sold in fluid form, and the rest was used in manufactured products.

FMMOs perform three major functions:

- Classify Grade-A milk according to how it is used—in fluid or manufacturing products. Fluid products are Class I use; "soft" products such as ice cream and cottage cheese are Class II use. Milk used in cheese production is Class III, and milk used to produce butter and dry milk powder is Class IV. Milk used in each of these different classes receives a different price.

- Set minimum prices for raw milk and milk components, with a higher price for fluid use (Class I) and lower prices for manufacturing uses (Classes II, III, and IV). FMMO staff audit milk processors to ensure that producers are paid these prices.

- Require that the proceeds from the sale of milk in a given federal milk market area be pooled among the producers supplying that market. All producers receive the average or blend value of milk sold based on the minimum prices established for each class of milk and the utilization of milk in the various classes.

There is a national market for dairy products such as butter, cheese, and NFDM because these products are storable and relatively cheap to transport. There is not much difference in the wholesale prices of these products in different parts of the country, and therefore, the price processors can afford to pay for milk to manufacture these

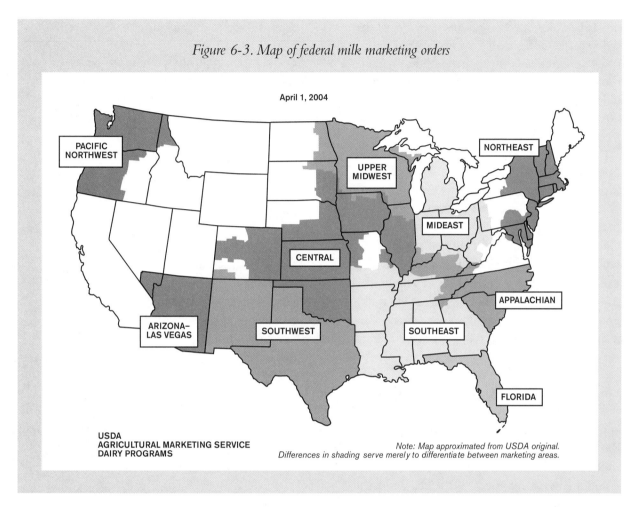

Figure 6-3. Map of federal milk marketing orders

April 1, 2004

PACIFIC NORTHWEST

NORTHEAST

UPPER MIDWEST

MIDEAST

CENTRAL

APPALACHIAN

ARIZONA-LAS VEGAS

SOUTHWEST

SOUTHEAST

FLORIDA

USDA
AGRICULTURAL MARKETING SERVICE
DAIRY PROGRAMS

Note: Map approximated from USDA original.
Differences in shading serve merely to differentiate between marketing areas.

products is similar around the country. FMMOs recognize this economic reality and set the same minimum Class II, III, and IV prices in the different order markets around the country.

The arguments made in support of the higher Class I price are as follows. Some additional costs are associated with the production of Grade-A milk. Fluid milk is normally processed and distributed in local or regional markets, because fluid milk is bulky, perishable, and expensive to transport, and there are certain costs associated with serving fluid markets.

Originally, FMMOs set the minimum Class I price in each market based on the price in Minnesota and Wisconsin (M–W). A market-specific Class I premium or "differential" was

added to the M–W price that, in general, reflected distance from Wisconsin and transportation cost. The logic was that Wisconsin, which had the largest supplies and a low Class I utilization, was the place to get additional milk when local supplies were short and the differentials were needed to encourage the movement of milk. So, for example, Florida had the highest differentials because it has a high Class I use and is farthest from the reserve supplies in Wisconsin. However, the FMMO Class I prices generated by this approach were above the costs of production in many areas of the country, which encouraged local production that was not needed to meet local Class I needs. These Class I differentials were not adjusted from the early 1960s until the 1985 farm bill, even though transportation costs had increased significantly. In 1985, because of general dissatisfaction with the regional structure of the differentials, Congress intervened and set new Class I differentials. After much political wrangling, the reforms adopted subsequent to the 1996 farm bill included relatively minor changes to regional Class I differentials.

From May 1995 until December 1999, the M-W price was replaced as the base price for setting FMMO class prices. The volume of manufacturing-grade milk had been declining for many years as producers in the two states either converted to Grade A or stopped producing milk altogether. The replacement, called the Basic Formula Price (BFP), was a rather complex formula based on wholesale dairy product prices for butter, cheese, and milk powder, weighted according to importance. The relationship between the M-W and BFP prices was very close, however.

On January 1, 2000, an entirely new FMMO price-setting mechanism was adopted. This new approach established the value of milk based on the wholesale price of cheese, butter, and nonfat milk solids. The value of the different milk components used in each FMMO class is derived from wholesale prices using product yield formulas and "make allowances" that reflect the cost of converting raw milk into specific dairy products. Producers are paid based on the quantity and value of the components in the milk they sell. Under these new procedures, Class II milk receives a small differential (premium) relative to milk used in Class III and Class IV. The new monthly Class I (fluid) price mover is the higher of the Class III or Class IV price, and Class I differentials are added to this. Producers in six of the 10 orders are paid on the pounds of milk components sold, not on milk volume. The remaining four orders, those with the highest Class I (fluid) utilization, pay on a per hundredweight basis. The method of payment in the different orders has implications for the choice of breed and the adoption of specific management practices that affect milk composition and value.

Federal orders regulate individual processors based on where they sell fluid products. If a processor sells in more than one FMMO area, that processor is regulated under the FMMO area with the largest volume of sales. Producers, in turn, are identified with, and paid under, the same federal order as the processor who buys their milk, regardless of the location of their farms. Therefore, neighbors may receive different prices for their milk because they sell to different processors, who are regulated under different federal orders.

The pooling provisions of FMMOs require all the milk handlers—farmer cooperatives and dairy processors—regulated under a particular order to share or pool their producer payments for the different classes of milk. Each producer then shares equally in the total sales for that federal order market, based on the minimum class prices in that order. Pooling does not affect what each processor pays for milk; it is a way of paying producers based on sales in the total market instead of the sales performance of an individual milk processor.

State regulations may affect milk marketing in place of, or in addition to, federal government involvement. The white areas shown in figure 6-3 (page 77) are not included in federal order markets. California, the largest milk-producing state, is the largest state-regulated milk market. In a survey conducted in the 1980s, 26 states had state milk marketing laws and regulations, but in the vast majority of cases the producers in these states also were regulated under federal orders. In 14 of the 26 states, there were laws and regulations covering trade practices but not milk prices. Twelve states had authority to establish minimum producer prices, five had authority to establish minimum retail prices, and 20 had authority to limit sales below cost. Eight states required price filing with a state agency.

The milk pricing provisions of state marketing orders are similar to federal orders in many respects and typically fulfill similar functions—for example, classifying milk based on use, setting legally binding minimum prices, and auditing handlers to ensure producers are paid as required by the order rules. The producer prices they establish are governed by the economic reality that milk and dairy products move across state lines, and prices in neighboring markets are linked by transportation costs. Therefore, the impact of the federal support price program and federal orders is felt even by producers who are not directly regulated under FMMOs.

As the previous discussion illustrates, government involvement in the U.S. dairy industry is continually changing. Two recent policy changes are noteworthy.

The Milk Income Loss Program was created by the 2002 farm bill. This is a new type of income support program for dairy farmers, with a three-year and ten-month life that extends until September 30, 2005. It is a counter-cyclical program that makes monthly payments whenever the reference price for milk falls below a specified target. This reference price is the Class I price in the Boston federal order market, and payments are made at a rate of 45% of the difference whenever the Boston price falls below the $16.94 per hundredweight target. Payments are made on milk sales up to a maximum of 2.4 million pounds per farm per fiscal year.

The intent of this program is to support dairy farmer incomes and to slow the rate of decline in dairy farm numbers. The payment cap is intended to direct more of the benefit to smaller farms. Relative to recent history, the $16.94 target is quite generous and frequent payments are likely. Payments were $1.08 per hundredweight, on average, during the first fiscal year. Program payments are expected to cause milk production

to be higher than otherwise would be the case, and therefore, farm prices will be lower. It is likely that periods of low prices will be extended, but not the periods of high prices. Lower market prices and the payment cap place large farms at a financial disadvantage, and this is likely to slow the persistent long-term trend to fewer but larger farms, but it will not reverse it. Also, because the proportion of large farms is higher in the West, there may be a slowing in the regional shifts in production documented in table 6-5 (page 71).

The Northeast Dairy Compact was a new type of regional dairy regulatory agency created during the 1990s. Six New England states created an interstate compact in response to low federal order prices and declining milk production and farm numbers. Any interstate compact must be created first by having the enabling legislation passed at the state level and then by obtaining authorization from the U.S. Congress. Congress approved the Northeast Compact in 1996, but this congressional authority expired on September 30, 2001. Fourteen southern states passed enabling legislation during the late 1990s for another compact, but Congress did not pass the necessary federal legislation.

The Northeast Compact had authority to set Class I prices in the six-state area, subject to certain safeguards. A commission administered the law, with representation from producers, processors, and consumers. Compact prices were minimum prices that had the force of law, and the commission had the authority to audit milk handlers to ensure producers receive the mandated prices.

Farm and Retail Price Relationship

The preceding discussion argued that supply and demand were the primary factors determining farm prices in the 1990s, and this is likely to continue. Figure 6-4 (page 80) illustrates the farm price relationships in different parts of the country, the balance between regional supply and demand, and

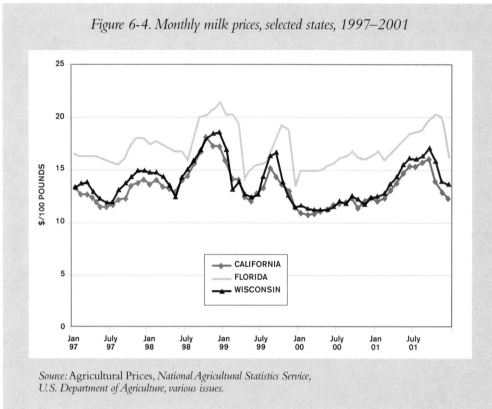

Figure 6-4. Monthly milk prices, selected states, 1997–2001

Source: Agricultural Prices, *National Agricultural Statistics Service,*
U.S. Department of Agriculture, various issues.

milk components, such as butterfat, or products, such as block cheese or milk powder, to another plant for further processing. The final product may be a recognizable dairy product, such as milk, butter, or cheese, or a prepared food with one or more dairy ingredients, such as milk powder in a bakery product or cheese on a pizza or hamburger. These products may pass through a wholesaler or chain store distribution center en route to the final retail outlet, which may sell products for consumption or meal preparation at home or away from home. Each step in the chain adds value in the form of processing, packaging, transportation, or other services. There is a corresponding cost to each of these types of added value.

classified pricing under federal and state marketing orders. California and Wisconsin are both well-suited to milk production, and a large proportion of the milk is used in the manufacture of dairy products. These products are relatively cheap to transport and are traded nationally. The resulting farm prices are similar and fluctuate with changes in the national supply and demand balance. Florida has high production costs, is a milk deficit region, and is far from alternative sources of supply. It has the highest farm price, but this price is also affected by the national supply and demand situation.

Milk passes through many hands on the way from the farm to the ultimate consumer. The milk truck that collects milk at the farm usually delivers it directly to the processing plant, but it may pass through a holding facility first. The processing plant may produce and package a product for direct delivery to a retail store, as is often the case for most fluid milk. However, the primary plant might sell

The difference between the price of a farm product and the retail price of the resulting food product depends on a number of factors, including the type of product; the amount of processing, packaging, and added convenience; and whether it is eaten at home or away from home. The USDA Economic Research Service estimated that in 2001 the farm value for a mix of dairy products was 34% of the final product value and the processing and marketing bill was 66%. The comparable numbers for a broad-based "market basket" of foods were 20% and 80%, respectively, in 2000. Over time, the farm share of the consumer dollar has fallen, because consumers have demanded—and have been willing to pay for—more and more value-added features, including packaging and convenience. For domestically produced food on the whole, almost half the marketing bill is labor cost, with packaging as the second largest item, at 8%. Advertising, energy, transportation, and profits each account for roughly 4% of the total.

According to USDA figures for 2001, American consumers spent 5.9% of their disposable income on food purchased for consumption at home and 4.0% on food eaten away from home. The total spent on food, at approximately 10% of disposable income, is the lowest in the world.

The farm price of milk is the result of a complex interplay of market forces. It changes in response to changes in milk production at the farm level, changes in consumer purchasing patterns, and changes in the costs of the various components of the marketing margin. Farm prices are extremely sensitive to fluctuations in raw milk production and changes in consumer demand. This price volatility is aggravated because retail prices are "sticky" and do not respond quickly or fully to changes in raw milk prices.

The markets for butter, cheese, and milk powder are national in scope because these products are storable and relatively cheap to transport. Raw milk and fluid milk products are bulky, perishable, and relatively expensive to transport, so markets for these products tend to be regional. Prices are higher in markets where a large proportion of milk goes into fluid use. However, these regional markets are still affected by the overall supply and demand balance because competition for milk and sales makes regional prices move in concert with national prices.

The Role of Cooperatives in Milk Marketing

Cooperatives market milk on behalf of their members, negotiate price premiums, and are active in the policy-making process at the federal and state levels. Most dairy farmers belong to cooperatives, but there is great diversity among cooperatives, including their size and the nature of the activities in which they engage. Some cooperatives are small as measured by the number of farmers or the volume of milk handled; others are very large. The largest dairy cooperative, Dairy Farmers of America (DFA), had more than 15,000 members in 2001. DFA members produced 34.5 billion pounds of milk, 21% of all the milk produced in the United States. Some cooperatives are involved in activities other than milk, but most specialize in milk. Some are involved in milk processing, while others simply try to find better paying markets or obtain a better price in the form of over-order premiums.

Traditionally, cooperatives have been a major force in milk marketing legislation and regulation. Cooperatives actively propose and monitor federal and state legislation that might affect their members. Involvement at the federal level includes changes in milk market orders, federal price and income support programs, the Dairy Export Incentive Program, and domestic feeding programs. Cooperatives also participate in the trade policy formulation and ratification process, such as the Uruguay Round, and monitor implementation. The level of involvement varies, however. Some are more active in the policy-making process than others. Many, but not all, cooperatives belong to the National Milk Producers Federation, the main political organization representing producer cooperatives, and some have their own policy agenda. The type and degree of government involvement in milk marketing owes much to the dairy cooperatives.

An over-order premium is the difference between the price a processor pays for milk and the minimum federal order price. These premiums are obtained by negotiations between processors and producer associations. Associations are usually large dairy farmer cooperatives or pricing federations, which consist of several cooperatives to bargain more effectively for higher prices. Over-order premiums are fairly common in FMMOs and have been as high as $2.00 per hundredweight for Class I milk. One justification given is that these premiums reflect additional marketing services provided by a cooperative to a processor. In some regions and time periods, a second justification offered is that the minimum federal order Class I price is too low to support adequate levels of local milk production but that, even with the over-order premium, local milk is still cheaper than milk imported from other parts of the country.

Over-order premiums, by definition, are not regulated or audited by FMMOs and are not part of the pool. Many factors affect the size of an over-order premium, including the availability of milk in a market area, the availability and cost of bringing milk in from sources outside the normal market supply area, and the size and strength of the producer pricing federation. To be effective, a pricing federation must control most of the milk in a large geographic area; 90% is a frequently cited target. However, the effectiveness of pricing federations is limited because of distrust between producer groups, pressure from individual processors on particular groups of producers, incentives for some producer groups to cheat on the negotiated prices, and lack of reliable information on prices paid in individual transactions. In spite of these weaknesses, the pricing federation concept is one of a very few practical means for producers to influence the price they receive for their milk other than through the political process.

Other Factors

Several other factors affect the price a farmer receives for milk:

- The farm value depends on the specific composition of the milk—for example, its butterfat content. The federal order reforms implemented on January 1, 2000, increased the emphasis on component pricing, although these new rules are more complex. The butterfat and nonfat solids in milk are affected by choice of breed, herd genetics, ration formulation, weather, and other factors.

- Milk hauling costs are borne by the farmer, usually as a deduction from the gross price. These charges are assessed on a per-hundredweight basis and may include a fixed charge per pickup. Charges vary by handler, and some subsidize this cost as an incentive for producers to sell to them. Large farms may be able to bargain for lower rates, and very large farms may find it more economical to own their own milk trucks.

- Some milk buyers offer quality premiums related to bacterial counts or related measures. These premiums reflect the benefit to processors of higher quality raw milk in the form of improved shelf life, reduced spoilage, or higher product yield.

- Some cooperatives offer premiums favoring larger farms, in part to attract or retain larger farms as members of the cooperative and in part to share the fixed costs associated with cooperative membership more equitably.

These premiums and deductions tend to be fairly stable, however, and do not contribute substantially to the large month-to-month variation in the prices farmers see in their milk checks.

Producers can also influence the prices they receive through price-risk management strategies. Dairy futures and options provide both producers and processors tools to help manage price risk. Also, some cooperatives and other milk buyers offer their members forward pricing contracts and other risk management opportunities. The milk price volatility that has occurred since the early 1990s stimulated the development of futures markets for several dairy products. These included cheese, butter, and nonfat dry milk. Later, futures contracts were developed for milk, including contracts for both the monthly Class III and Class IV prices. In addition to hedging opportunities, futures prices provide useful information about price expectations for near-term business decisions.

Selecting a Milk Buyer

As discussed above, a variety of market factors affect the price received by producers. The combination of these factors may cause one farmer to receive a price that is markedly different from the price received by a neighbor. For these and other reasons, a producer should consider a number of factors when selecting a milk buyer. These factors can be classified under philosophy, price, and market security.

Producers may have a choice between selling their milk directly to a milk processor as an independent producer or belonging to and selling through a cooperative.

Although price and market security are important considerations, many producers also make decisions based on their philosophy about, attitude toward, and experiences with marketing their farm products cooperatively.

A producer's choice is also based on an assessment of the characteristics of individual cooperatives. There is great diversity among cooperatives, including their size and the nature of the activities in which they engage. Some are involved in processing; some simply try to find better-paying markets; some are more active in the policy-making process; some provide field services, selling

dairy supplies and insurance; and some offer forward pricing contracts as a price-risk management tool for members. Cooperative activities have a cost; some of these generate benefits for the cooperative members, and some generate benefits for members and nonmembers alike, so equity is a consideration. Individual producers must assess the relative merits when selecting a buyer.

As a matter of economic necessity, producers must be concerned about the mailbox price (the price a dairy farmer actually receives for milk) received for their milk. Market forces, the support price, federal and state orders, over-order premiums, and season affect all processors and producers in a similar way. However, two neighbors can receive a different mailbox price for their milk. In some cases, these differences are short-lived; in other cases, significant price differences can persist for months or even years.

Mailbox price differences can occur for various reasons or a combination of reasons. Milk might be sold in different markets with different prices and different Class I utilization rates. For example, one farmer might sell his milk to a handler regulated under a federal order, and the other might sell to a plant regulated by a state agency. There could be differences in milk components, such as butterfat and protein. In this situation, prices should be compared on a standardized basis to identify true price differences. Hauling charges or quality and volume premiums may differ. One producer might be a cooperative member and the other might not, or producers might belong to different cooperatives that sold their members' milk in different markets or that had different operating costs.

The issue of cooperative pay prices requires some additional explanation. All cooperatives have operating expenses, and these must be paid from the money received by the cooperative from the sale of members' milk. These operating expenses and deductions vary, depending on the type of cooperative, the activities in which it engages, and the efficiency with which it operates. Cooperative members provide working capital through capital retains programs; a small part of the farmers' income is withheld and used to supply the cooperative with working capital. In addition, some retain a portion of the profits (patronage dividends) that would otherwise be returned to the members. Members are denied the use of this money until it revolves out of the working capital fund after a period of years. As a consequence of these factors, cooperative members often receive a lower monthly price than producers who do not belong to cooperatives, but this is not a fair comparison unless the other benefits are considered, such as the annual patronage dividends (profits) paid to members, the value of any field services, the lower cost of supplies and services sold to members, and market security.

Bankruptcy and other financial difficulties among major U.S. corporations are commonly seen headlines. These events are uncommon among dairy processors, but there have been instances in which processors ceased operation and failed to pay producers for milk. Producers are normally classified as unsecured creditors and receive no or only partial payment of what they are owed, often after a considerable delay. Independent producers bear the full amount of their individual loss. Most cooperatives supply milk to more than one processor and, therefore, are able to distribute the losses among the membership at large. Plant closings, whether through bankruptcy, takeover, merger, or restructuring, can also force independent producers to seek another buyer, often at short notice, whereas cooperatives are better placed to find alternative buyers for the displaced milk. Bankruptcy, takeovers, and mergers may also deny producers other services provided by their milk buyers, such as insurance and retirement programs. Takeovers and mergers among both cooperatives and processors occurred

at a fairly rapid pace in the 1990s, and producers are well advised to monitor trends and investigate the viability of milk buyers.

Summary: Marketing Milk as a Commodity

The United States is largely self-sufficient in milk production. Domestic prices normally are significantly higher than prevailing world prices, which are artificially low because of subsidized exports from the European Union. This limits export opportunities. Imports are restricted by tariff-rate quotas. No new trade agreements are imminent, and the U.S. dairy industry will continue to be affected primarily by domestic economic forces and policies for the foreseeable future.

Domestic consumption of dairy products increased slowly during the 1980–2001 period, approximately 2.0% per year on average. Milk production per cow has increased more rapidly than sales, 2.5% per year, and these production increases have put downward pressure on farm prices. Cow numbers have declined as a result of this price pressure and other factors, which have kept production and sales in relatively close balance. Farm milk prices were heavily influenced by the federal price support program prior to 1989, but "market forces" took over during the 1989–2001 period. A reasonably close balance between production and sales, combined with seasonal factors, caused extreme month-to-month price volatility.

Prior to 1989 the federal dairy price support program was an important factor in setting milk prices across the nation, but its importance has declined because the support price has been reduced to levels that are considerably below the average cost of production. Nevertheless, it still provides a floor or safety net, albeit at a low level.

The federal milk market order system regulates most of the Grade-A milk produced in the United States. Federal orders used the manufacturing milk price as the base point for setting minimum farm prices until December 1999. Under order reform implemented in January 2000, federal order minimum prices are based on wholesale prices for dairy products. The value of the milk is derived using yield formulas and make allowances. Since 1989, for the most part, wholesale dairy prices have been set by market forces, but the support price has had a sporadic impact, largely in times of seasonal surplus.

Federal orders establish higher prices for milk used in fluid (Class I) products. Class I differentials are related in part to distance from Wisconsin. They are smallest in the West and Upper Midwest and are largest in the Northeast and Southeast. However, in many markets these minimum prices are considered to be too low to support the desired volumes of local milk. During the last 20 years, the larger cooperatives and, more recently, producer pricing federations, have been successful in negotiating over-order premiums on Class I milk. The size of this premium depends on the availability of milk and the strength of the pricing federation.

The combination of national supply and demand conditions, the price support program, and the federal order system affects prices even in areas under state regulation because prices cannot stray too far out of line with neighboring markets.

All of these factors—market forces, the support price, federal orders, over-order premiums, and season—affect all processors and producers in a similar way. However, they do not explain why two neighbors receive a different mailbox price for their milk. These differences can occur for various reasons or for a combination of reasons, including: (1) milk is sold in different markets with different prices and a different Class I utilization; (2) milk is sold by cooperative members or an independent producer, or is sold through different cooperatives selling milk in different markets or having different operating costs; (3) there are differences in milk components,

hauling charges, and quality and volume premiums. The chain of events leading to a producer's mailbox money is depicted in figure 6-5 (page 86).

One final and very important point: A study of dairy farm financial records shows a large variation in financial performance among dairy farms. However, differences in the price received for milk explain relatively little of this variation. Other important factors include differences in animal performance and cost of production. Ultimately, the responsibility for financial performance lies with the primary operator(s), so it may be fair to say that the most important factor is level of management skill and ability. The price of milk alone does not explain why one dairy farm fails while another survives and prospers.

VALUE-ADDED PROCESSING AND MARKETING

Consumer interest in specialty dairy products seems to be increasing, though hard numbers are scarce. Specialty cows' milk and goats' milk products are sold in a variety of forms and through a number of different market channels. Some consumers are looking for items not generally available in supermarkets, such as specialty cheeses or items identified as local farm products. Other consumers are concerned about how food is produced and are looking for items produced in a certain way or under specified conditions, such as certified organic products or products from animals raised on pasture instead of in confinement. Examples include farm-bottled milk, cheese, ice cream, yogurt, and special products for specific ethnic markets. These products may be sold directly to consumers at a farm store or farmers market, they may be delivered to a retail establishment for sale, or they may be sold to a wholesaler for resale at a retail establishment. Some of these opportunities may be small-scale and local, but as the market success of Ben and Jerry's ice cream and Horizon Dairy's organic products shows,

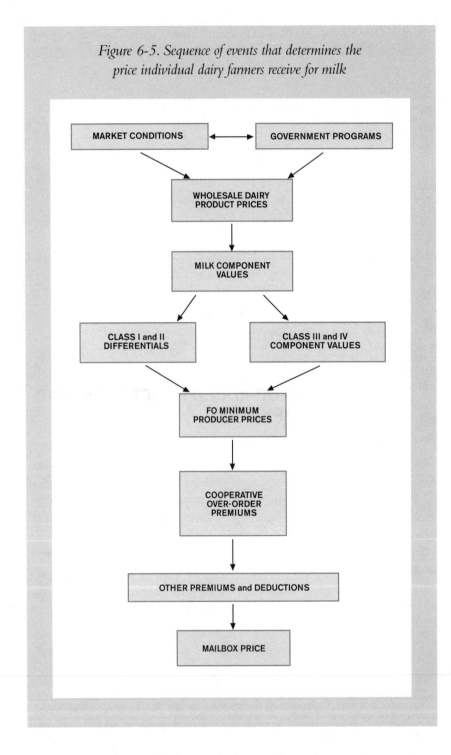

Figure 6-5. Sequence of events that determines the price individual dairy farmers receive for milk

MARKET CONDITIONS ↔ GOVERNMENT PROGRAMS

WHOLESALE DAIRY PRODUCT PRICES

MILK COMPONENT VALUES

CLASS I and II DIFFERENTIALS

CLASS III and IV COMPONENT VALUES

FO MINIMUM PRODUCER PRICES

COOPERATIVE OVER-ORDER PREMIUMS

OTHER PREMIUMS and DEDUCTIONS

MAILBOX PRICE

products, these alternative markets can provide business opportunities for people with the necessary desire and talent. On the plus side, these specialty or niche markets can attract consumers who are willing and can afford to pay a premium price for the products they want. On the negative side, competition for consumers' time, attention, and dollars is fierce. Many of these opportunities are local and unique and, therefore, anyone considering dairy niche market opportunities should conduct their own assessment before making any commitments. There is no substitute for developing a comprehensive business plan for the specific farm and family situation. Developing a business plan takes time and effort, but it is both necessary and beneficial when one considers the risk involved and the time, effort, and financial resources invested over the life of a new business. In this respect, dairy niche markets are no different from any other business venture.

consumers' needs may also be met by large-scale businesses that operate regionally or nationally.

Although the overall level of consumer interest may be small relative to the total market for dairy

Much has been written about the process of business planning and, as a key component, developing a marketing plan. Information on the subject is readily available in many public libraries and through the Internet. Assistance is available from a variety of sources, including land-grant university extension services, community colleges, state government agencies, and consultants. The main focus of the federal government's Small Business Administration is nonfarm business, but it has educational materials, training programs, and resource links. Producers will have to sift through these resources to find those most closely related to the proposed type of business. The following discussion provides a brief overview of business planning, including developing a marketing plan.

A business plan is really just the carefully thought out answers to some very important questions:

1. What are our goals, including financial and lifestyle goals?

2. What should the business look like to accomplish these goals?

3. Who is the competition and how do we stack up?

4. What will we produce and how and where will we produce it? What governmental or other rules and regulations must we comply with?

5. Where and how will we sell our products, what will the prices be, and how will we convince customers to buy them?

6. What resources will we need, including financial investments, working capital, people, management skills, materials, and services?

7. What do we project for income, expenses, profits, and cash flow?

The business plan should be written, because this will help ensure that it is complete and that all the essential people fully understand what is being proposed, and because it will help in obtaining financing. Developing a written business plan will not guarantee success, but it will certainly increase the chances. There is no standard format, but a written business plan is generally organized around the seven questions listed above.

The first question is the most important, because it provides the justification for the business venture and a sense of direction. It also forms the basis for choosing among alternatives and making decisions. Creating a new business or making major changes in an existing business is hard work and stressful, so it is important that the principals are in agreement. Regardless of the size of the proposed venture, all the key players should be involved in goal setting, whether they are family members, unrelated individuals, key employees, or a combination of these. These goals will typically include both financial and lifestyle considerations. Each individual should consider his or her personal goals and those of other family members to make sure they are consistent with the goals of the business.

Many businesses have found it helpful to consider broad, long-term objectives (say, 10 years out) and more specific short-term goals that will lead incrementally to the achievement of the long-term objectives. Some businesses find it helpful to distill the key goals, values, and business philosophy into a mission statement. This mission statement serves as a reminder of what the business is all about and helps communicate this to employees, suppliers, and customers.

To meet the goals of the owners, a business must be able to meet customers' needs profitably. Questions 2 through 5 relate to different aspects of designing a profitable business. The answers to these questions usually reflect a trade-off between the interests, preferences, and quality of life goals of the owners on the one hand and the economic realities of the marketplace on the other. Therefore, when developing a business plan, the fifth question may be the second most important, and answering it requires the development of a marketing plan.

A successful business is built on a base of satisfied customers. The marketing plan involves identifying customer needs, assessing the competitive advantages and disadvantages of the proposed business relative to other competitors, choosing the target group(s) of customers, and figuring out how to meet the specific needs of these customers and convince them to buy the product(s). At the most basic level, this involves the four "Ps"— product, place, price, and promotion. Product refers to the specific features of the items offered for sale, including type, size, packaging, and services. Place describes how to get the product to the location where the consumer will buy it, including the market channel(s), transportation, and use of middlemen. Price considerations include the cost of production and marketing, the prices charged by competing suppliers of similar products, the prices of competing products, and the relationship between the price level and the total quantity customers are willing to buy (price elasticity). Promotion is how potential customers find out about the special features, advantages, or benefits of the product(s) and are persuaded to buy. These questions cannot be answered without valid information gathered through market research.

Market research is the basis for a sound marketing plan. This research helps identify potential customers, including what characteristics they value and want in a product, how many customers there are, where they are located, what their buying habits are, and so forth. When creating a market for a new product, this information will help you develop product specifications, make sales and income projections, and decide which specific market channels and locations are most appropriate. When planning to sell into an existing market, information about the total size of the market and current trends is important. Is the market growing, shrinking, or stable? Who are the competition, and what are their relative strengths and weaknesses? Realistically, how much of the market can be captured and by what means? How are the other sellers in this market likely to react if a new competitor enters the market?

One important decision is whether to sell directly to consumers, to sell indirectly through a middleman, or to use both options. If you sell direct, you must gather all the needed marketing information yourself. If you sell through a middleman, that middleman can provide a great deal of information and will often make suggestions or impose requirements related to product type, volume, packaging, price, and so on. The choice will depend partly on your goals, partly on your skills, and partly on the expected costs and returns from selling through different market channels.

The information generated by market research, along with a large dose of creativity, will guide the development of the specific marketing strategy to convince customers to buy the proposed new product(s). However, markets change continually, and there is a corresponding need for continuous market research.

A great deal of information is readily available at low or no cost from government, industry, or other sources. This information includes consumer demographics and buying habits, industry and economic trends, and so forth. However, these secondary sources of information often do not provide the specific and detailed information needed for a marketing plan. Therefore, additional research likely will be needed that involves collecting information directly from potential customers using personal or telephone interviews or mailed questionnaires. Focus groups and taste panels may be helpful for new food products. Great care must be taken in designing surveys, including which questions to ask, how to ask the questions, how many people to contact, and how the data are to be analyzed and interpreted. Gathering these types of data takes time and money, but there are few alternatives if the marketing plan is to have a sound factual base. Many farm families lack expertise in marketing and market research and may benefit from the services of a consultant. The information in the marketing part of the business plan can then be used to develop or revise the production part of the plan, question 4 in the previous list.

A number of regulations apply to milk production and processing, and these must be identified as part of the planning process. Applicable federal, state, and local regulations may include zoning, sanitation and health, bonding, licensing, weights and standards, labeling, environment, waste management, and federal or state milk orders. Some of these regulations involve prior approval, some impose specifications on facility construction and operating practices, and some involve routine inspections or quality-control measures. Organic standards and certification are regulated under the National Organic Program, as authorized under the Organic Foods Production Act of 1990. Organic certification is handled by federally approved private agencies and some state departments of agriculture. Compliance with the myriad rules and regulations makes demands on time and energy and adds to the cost and frustration of being a small business operator.

Milk production is likely to vary seasonally, and product sales may fluctuate throughout the year. The marketing and production plan must recognize potential imbalances and develop a strategy to cope with them. This may include producing a mix of products, producing some storable products, or finding a market for any surplus raw milk. The production and marketing plans, in turn, are used to identify the resources needed to implement the plan, question 6.

The production and marketing plans must be in harmony. The market opportunities help define the type and capacity of the facilities, processing and distribution equipment, storage, supplies, and labor needs. Some flexibility in the production process is desirable to allow adjustments in product

mix and volumes in response to actual sales experience. Facilities and equipment should be designed with potential expansion opportunities in mind. An allowance for cost overruns and unanticipated problems should be built into the facilities and equipment budget. Estimates of costs of production are essential in selecting among alternative production systems, and unit production costs are an important consideration in pricing and, therefore, in expected sales volume and profit potential.

The last question pulls all of the information together to assess the financial performance expected from the business. All new ventures go through a startup phase, and there may be significant up-front investments and expenditures before any sizeable revenues are generated. This is particularly true for new organic production systems, which have a mandatory three-year transition period. Therefore, projections of financial performance should begin when the first financial resources are committed, extend through this startup period, and include the projections for the fully established business.

Profit measures include all revenue and costs and account for noncash items such as inventory increases and depreciation charges related to the investment in business assets. Cash flow projections are needed to evaluate the timing of cash inflows and outflows in order to assess credit needs and establish reasonable debt repayment schedules. These cash flow projections should be made on a monthly or quarterly basis and should extend through the startup phase to the point at which the business is in full operation. The money available for the owners to withdraw from the business to meet personal and family financial needs is an important component of cash flow feasibility for a family-owned business. Measures of financial performance are very helpful in assessing the viability of the plan. The rate of return on investment is one key profit measure. The ratio of income to expenses, debt-to-asset ratios, and debt repayment capacity are other important performance measures.

Because any business faces a number of risks, profit and cash flow projections should be evaluated under a variety of production and sales assumptions to see how well the business would fare and to develop contingency plans. If the initial projections are unfavorable, all the effort that went into the planning process has still been worthwhile. It may have prevented a major disaster and should be taken as a challenge to go back to the drawing board and develop an alternative plan.

Once a viable business plan has been developed and the necessary financing obtained, the business plan should be kept and used as the blueprint during the implementation phase. If unforeseen circumstances arise and it is necessary to deviate from the plan, figure the financial consequences before making any changes. It is important to monitor actual investments, production and sales, and expenses and revenues by comparing them to the projections. This provides early warning of production, sales, and financial problems. An effective monitoring system also provides the necessary information for making decisions and taking corrective action promptly.

CONCLUSION

The U.S. dairy industry is dynamic and will continue to be so. Producers need to be alert to changes that might affect them and must respond as the economic environment changes. Several trends are likely to continue. Consolidation will continue among processing firms and cooperatives. Government policies will continue to have a significant influence on the U.S. dairy industry. Trade liberalization is likely to continue, albeit slowly. The precise nature and timing of trade policies affecting dairy products will determine whether U.S. producers will benefit or be harmed. The U.S. Congress seems to have abandoned the "Freedom to Farm" concept in favor of safety nets, at least for the traditional farm commodities. Consumer concerns about the health and safety of food products seem to be growing, as are concerns about farm production practices. Food consumption patterns will continue to change, along with changes in the racial mix, age, income, and lifestyle demographics. All of these changes and some that cannot be foreseen will affect milk production and dairy product marketing. For some, they represent threats; for others, they are opportunities.

CHAPTER 7
Direct Marketing

Thomas R. McConnell

Direct marketing a product is about the only time a farmer has the opportunity to be a price maker. Probably all grass-based farmers consider direct marketing. After a trip through the meat or dairy section of the local market, or when they learn what their neighbor paid for wool yarn, they wonder: Why can't the farmer get some of that mark-up? Some farmers take that first step into direct marketing by connecting with a neighbor or family member for a side of beef or lamb. That experience can begin an enterprise that provides direct consumer feedback and the personal satisfaction and sense of community one gets by feeding one's neighbors and friends wholesome, fairly priced farm products. For others, that experience may indicate that direct marketing has many more obstacles than they care to overcome and requires more time than they feel they have.

Marketing directly means circumventing the standard channels or steps in processing and commerce between the farm and the end-user of a product. From the direct marketer's perspective, the buyer may or may not be the consumer. Direct marketers and their customers differ in the amount of the marketing they choose to perform themselves. In the case of meat animals, the producer may arrange the sale, provide the transportation, oversee the slaughter and processing, and, most importantly, price the product. Not all direct marketers are involved in every step from production to consumption. In some cases, the buyers assume more of the responsibility for other steps in the process, and in the case of meat animals, this might include processing. Some buyers insist on doing the slaughter. One successful direct marketer sells nearly every lamb directly off the farm, with his

scales and his price. A goat cheese farm family uses a combination of the Internet, knocking on doors, and repeat customers to move all their products. This family's skills include being farmers first, cheese makers second, and direct marketers last. Their consumers perform no other task but paying for and eating the product. Some shepherds direct market their wool and hair to spinners and felters using the Internet as their marketplace. Others add more value by selling their wool as blankets, garments, and yarn.

WHY DIRECT MARKET?

Direct marketing changes the traditional role farmers play as price takers to one in which they set the price for their product. Since there is really no other product like it or reference price—in the case of hand-spun wool or the recently legitimized and touted dairy products and meat from grass-fed animals—***direct marketing allows the producer to establish a previously undetermined market and price***.

This discussion will concentrate on the decision to market beef, lamb, goat meat, dairy products, and fiber directly to a consumer. Small markets, restaurants, and processors are included in the consumer group.

Farmers market directly for many reasons. The anticipated increased net income from their product is one reason. This is viewed by very small and startup operations as a means to trade volume for margin. Some farmers direct market to gain more information about their breeding program as it relates to carcass quality and usefulness. The face-to-face interaction with the consumer attracts many; compensation includes more than money. Most direct marketers either enjoy working with the public or disguise their dislike very well. The farmers who don't like to interact with the public give that reason for quitting. Other reasons for quitting include the time required to adequately manage a direct marketing operation; mistakenly choosing the direct marketing outlet for inferior or ill-managed products or for those products that don't fit the market because of conformation or usefulness; and the wait for payment after a sale.

Direct marketing is a high-energy, intensive management option. Small farmers choose this option more often than the managers of larger operations. Small farmers feel they have more time to carry out the many tasks required to successfully direct market than do the managers of large operations. Successful direct marketers are effective managers, and effective management requires time. Most large operators choose to expend their management resources in other areas of their businesses. There are some very successful large operations that market directly to the end user or distributor. These operations employ expert managers to handle every task the individual farmer does himself. Often these operations succumb to the temptation to turn production over to their neighbors and concentrate on further processing and marketing. The small farmer, who often wants to increase his or her farm income, will direct market to increase revenue without increasing production volume. Many farmers who have off-farm jobs can look toward their coworkers as potential customers. Some farms that sell vegetables off the farm often add meat as another product to offer.

Experience proves that an operation's commitment to direct marketing is more a function of dependence on the income derived from it than the time spent doing it. The best position from which to start a direct marketing

enterprise is with a fresh, high-quality product, whether it is a meat animal; a clean, easily spun fleece; or a wholesome dairy product.

WHAT DOES DIRECT MARKETING INVOLVE?

The greatest change farmers report after making the transition to direct marketing is an increased time commitment. Lack of time is the most commonly reported reason bad decisions are made and good ones are not implemented. As the quick budget for hauling on page 94 shows, the value of time can have important implications regarding the cost per unit of any product.

The direct marketer's list of tasks may include:

- Market research
- Advertising
- Making a sale
- Sorting/selecting and delivering the products
- Making arrangements with the slaughterhouse
- Customer coordination regarding processing, pickup, and delivery
- Collecting money
- Customer relations and handling complaints

Market Research

The potential direct marketer should first research the market to determine if there is opportunity to sell his or her product directly. Research should start, at the very least, with a call to an extension office and continue with a visit to the local farmers market. The impression nationally is that the number of farmers entering the direct market movement is increasing.

The direct-market customer also has a list of reasons to buy directly. These might include issues such as freshness and quality, price, support of the farm community that protects the environment, and their disposable income. A 2002 report revealed that sweet corn at a Martha's Vineyard farmers market was selling for $9 per dozen, while the same product in a very rural market in West Virginia was moving at $2.50. So the decision to market directly should be made only after it has been determined that there is a sufficient customer base to support the seller at a reasonable price. Prospective direct marketers can be severely misled by interpreting out of context information gleaned from presentations by successful contemporaries. Farmers who market on the urban/rural fringe, or those who truck their farm production into the city, have a much brighter perspective than those who produce and sell in sparsely populated areas. Those reporting the universal success of direct marketing often hold a "one size fits all" mentality.

Advertising

The beginning direct marketer should consider a comprehensive advertising program—one that is not necessarily expensive, but one that speaks to more people than he or she sees every day. The marketer should keep his prospective buyers in mind in his advertising campaign. Yet most direct marketers start out with word-of-mouth, brochures, and newspaper advertisements. It is logical to think that the normal advertising channels used by suppliers of mass market products also will be effective for the grass-roots direct marketer. The advertisement should explain the benefits of the grass-roots influence on the product and the importance of the family farm community. The advertising rule might be: the more discerning the targeted customer, the more specific and focused the advertisements should be. Producers of gourmet cheese would want to target the segment of the population that appreciates it and, of course, can afford it. Or the opposite may be true: the more mainstream the product, the more conventional the advertising campaign should be. These may even include

newspaper advertisements. This type of advertisement has widespread coverage and will generate many inquiries or begin to reveal to the marketer just how strong the demand is. The responses will vary; they can yield many "unfertile" responses. Many direct marketers of meat have acknowledged that a broad advertising campaign resulted in a response from the local health official about how they were processing this meat. Another call might come from a local animal welfare group that probably is not as enthusiastic about your enterprise as others are. From the management point of view, the advertisement will help focus the marketing plan to make transportation and slaughter more efficient. Many direct marketers strongly recommend an advertising plan describing your product and what processing and marketing protocol you would like to follow. Many small direct marketers find their customers are satisfied with a seasonal marketing period. The steady customers appreciate receiving postcards that arrive near the time the product is ready, because it frees them from having to remember to call the farmer.

Making a Sale

Making a sale involves more time than most direct marketers anticipate. The personal touch and the attention to detail associated with a direct sale will allow the buyer to connect to the farm and the farm family's way of life. This same personal touch can become very time-consuming and rob the farmer of time needed to harvest or time spent with family. In a successful direct on-farm sale enterprise, the buyer is clearly informed about the price, the yield, and the terms for payment. Many first-time customers are overwhelmed by an invoice for a whole beef. They also are not aware of the amount of freezer space required to hold a beef. Hand spinners often confuse a per-pound price with the price for a whole fleece in the grease. Good communication will prevent surprises and ensure that the buyer is ready to make this commitment.

Determining Direct Marketing Expenses

Those who direct market meat and milk must pay careful attention to transportation expenses. Getting an animal to the slaughterhouse can be as simple as penning one off and waiting for a commercial trucker or neighbor to haul it for you. It can also include catching the entire herd, selecting an animal, separating it from the herd, and hauling it several miles or paying an employee to haul it (or covering for a farmer while he or she hauls it). This can be very costly, and it demands a lot of coordination to eliminate inefficiency. Many direct marketers report dissatisfaction with this option because of the time and money involved. Small ruminants, on a per-head basis, can be very expensive to haul. The mistake that many eager marketers make involves taking that *one* lamb or kid to the slaughterhouse to make a deal and get a new customer. A huge loss per head on the first lamb could require many years to overcome, as the budget below demonstrates.

These are realistic examples, though every marketer will have a different experience and situation. The point is to make budget determinations before the sale is made. As local slaughterhouses continue to disappear, it becomes obvious that trucking efficiency in direct marketing is necessary. Great inefficiencies can sneak up on a busy, overextended farmer and rob him or her of much deserved profit.

Getting Paid

How the product was raised, processed, or marketed will matter very little if the farmer doesn't get paid. When the meat and the buyer are at the slaughterhouse and the farmer is not, payment will not be a priority. Careful attention to how and when the product will be paid for is very important and must be settled early in the negotiations. Customer satisfaction can often get mixed up with money collection, requiring more time and effort than most beginning marketers estimate. Many direct marketers have experienced great frustration at either waiting for money or not getting paid.

Quick Budget for Hauling

OPTION 1: Two lambs weighing 85 pounds each hauled 50 miles to the slaughter facility and return trip:

$$(50 \text{ miles} \times \$0.50/\text{mile}) = \$25.00$$
$$\text{Return trip} \qquad \times 2$$
$$\$50.00$$
$$\text{Hired help, 3 hours @ \$7.00/hour} \quad \$21.00$$
$$\$71.00/2 = \$35.50/\text{head}$$

Spread over two 85-pound lambs = $35.50/85 pounds = $0.41/pound live

And consider dressing 50% $35.50/42.5 pounds = $0.84/pound is added to the carcass weight price

OPTION 2: One 1,100-pound steer that dresses 62% would be a much more efficient haul

$71.00 (mileage and labor)/1,100 pounds = $0.065/pound live

Consider dressing 62% $71.00/682 = $0.10/pound added to the carcass weight price

---

Attaining customer satisfaction can be the most important part of the direct marketer's work. When a customer doesn't understand the product and how to use it, doesn't understand his or her bill, or is allowed to fret or stew over an easily corrected but ignored issue, repeat sales will be difficult to make. Good direct marketers set aside time for follow-up calls to customers to focus on these details.

COMMUNITY AND SALES

Location plays a significant role in the success or failure of your direct marketing effort. This is a very simple concept. The size and proximity of the community to which farmers choose to market will influence the number of potential customers. The composition of the community, its income level, and ethnicity may help predict the potential for sales. Researching the community at the local chamber of commerce and reading the local paper will help the you explore the potential market. This type of information can serve to persuade a potential marketer to look to settle in a particular community, but most farmers buy their farms first and then decide to direct market. Proximity of the farm to the community will play a role in the feeling of connection the buyer has with the farm. Many direct marketers are amazed by the sense of connection their customers have toward the farm and the sense of partnership that evolves between the farmer and the customer. This can be a source of comfort for certain marketers or another reason to quit direct marketing for others.

NICHE OPPORTUNITIES

Many marketers who previously thought they didn't have the location to market directly have turned to a Web-based market presence. Web sites can be very well written, well designed, and well managed to portray a very positive and/or romantic image of the family farm attempting to

sell a product. The Internet affords the farmer the opportunity to craft precisely the market image desired. The Web site can provide the potential customer with a consistent message and farm philosophy. The site can help develop the bond between the customer and the producer. Some marketers are enjoying great success with Internet-based marketing.

Many farmers report increased demand for their grass-fed meat and milk because their customers associate these products with increased levels of conjugated linoleic acid (CLA). CLA is a fatty acid that is naturally found at low levels in ruminant products. Studies show that feeding this fatty acid to rodents and other animal models reduces risk factors for several cancers and cardiovascular disease. Although there have not been studies investigating the direct effect of CLA on human health, epidemiological studies have shown a relationship between higher ruminant product consumption and reduced disease incidences. This discovery has encouraged many farmers to promote their products as health-enhancing food and will remove pasture-based red meat and milk from the general red meat category that consumers were previously warned to avoid or limit in their diets. When certain foods, such as pasture, plant oils, and fish oil, are fed to ruminants, CLA levels increase in meat and milk. This is an exciting prospect, but pasture-based producers must not think that this is the only way to increase the levels of this compound.

In addition to pasture and dietary oils, current research is investigating methods to enhance CLA in ruminant products by supplementing the precursor of CLA and increasing the activity of the enzyme that synthesizes it. CLA can be produced synthetically from linoleic acid and is available as a supplement. It is logical to assume that if there is a way to derive the "pasture-like" benefits of this compound in an "artificial" environment, then that is what the industry will do, and the pasture-produced advantage could be at risk. This will be determined in the future, but it is feasible that this small producer advantage could go the way of the organic advantage, when large agribusinesses adjusted their production protocol to qualify them to make organic food more mainstream. Thus, organic food is no longer strictly associated with small family farms.

Specialty Meats

The annual per-capita consumption of beef is nearly 100 pounds; consumption of lamb is approximately 1 pound, and goat is less than that. Therefore, shepherds must understand their market. When it comes to lambs, it seems the industries are parallel. There is the traditional commodity side in which most producers have operated for many years. This industry has satisfied America's diminishing "white tablecloth" and supermarket lamb market for decades. Then there is the "other" side of the industry that satisfies a growing ethnic market, including the demand for goats, in which the animals should be treated as a specialty item. Understanding that there is a two-market system should encourage the shepherd to learn about it. The two-market system also can change pricing, because there is a very strong seasonal or holiday demand. Though the commodity side of the industry views seasonal availability as a problem for steady lamb prices, the specialty side celebrates the insatiable demand during religious holidays. Direct marketers need to understand this difference, and they need to produce and market accordingly.

Direct marketers must know the significance of the Christian, Islamic, and Jewish holidays and celebrations. The list includes, but is not limited to, Easter, Christmas, Eid al Fitr, Eid ul Adha, Passover, Rosh Hashanah, Hanukkah, and Ramadan. Most importantly, they need to know how these holidays relate to lamb and goat consumption. This knowledge will affect the timing of the breeding of their does and ewes. Producers must know the particular ethnic preferences relating to lamb and kid weight, castrating and docking, and market times. Many buyers will demand provisions made for strict adherence to kosher or Hallal ritual. Locally, there is no point in guessing; the aggressive direct marketer will contact the local leadership of these ethnic communities for specific information. Many marketers report that they have established very solid relationships with very good customers using this tactic. Many direct/niche marketers have chosen to offer on-farm slaughter for their customers, though some provision has to be made for disposal of the offal.

Lambs and kids possess great flexibility for carcass suitability. Many buyers want them at very low weights; for example, capretto kid or suckling kid carcass weights are acceptable as low as 13–15 pounds. These low weights bring very high prices per pound. The producer must consider whether the high price per pound equals an acceptable total gross receipt per ewe.

Overall with lamb and kid, marketers find that their new or potential buyers don't know how to prepare the meat. It is a good idea to arm these newcomers and many established customers with recipes from a Web site or of your own, or to put them in touch with a potential mentor.

PRICING

No single issue is more difficult than establishing a price. If you add no value to your product, your decision is easy. The producer may price at a respectable margin above local market quotes or perhaps establish some steady price that provides a comfortable margin. If you are carrying out many of the steps required to get the animal from the farm to the customer, you must assign a base value and determine a cost for each step of integration. Guessing or borrowing a price from your neighbor is one way to get started and may be one

Direct Market Budget Form (financial/time)

Activity	Cost	Time
Advertising	$	hours
Personal contact	$	hours
Selecting/delivering animals	$	hours
Processing	$	hours
Billing/collection	$	hours
Follow up/customer satisfaction	$	hours
Subtotal	**$**	**hours**

Value of labor in $ x hours _____

Total = cost + labor cost = total $_____

Total $/total pounds marketed _____ = Cost per pound marketed _____

Cost per unit expense live weight _____

Cost per unit direct marketing costs _____

Total cost per pound marketed (live weight) _____

way to get out of business very quickly. The central principle regarding pricing *your* product is to determine what *you* have to receive for the labor and investment *you* have made in producing it. The recurring theme is a personal one. No one else can price your product. Unfortunately, farmers aren't good at this, because they have been price takers for so long.

The first essential step in establishing a price for a product is to determine what it costs to produce it. Start with an accurate cost per pound of production. Allocating your chosen enterprise against the total farm records reported on a Schedule F form from the Internal Revenue Service will provide a very good start. The key is to be thorough and complete. If a direct marketing budget does not include both labor and risk, the whole price will be built on a weak foundation. Then, if the customer refuses the price the farmer has suggested, the farmer will not have the confidence to hold firm. Refusing to take an

offer or having a customer refuse a farmer's price is not the same as losing a sale. It only means that the farmer and buyer could not agree. But the farmer can feel confident in the decision if it is based on a thorough cost analysis.

After an accurate cost per unit of production has been established, the marketer must account for every single task that needs to be completed before the product is delivered to the consumer. Some tasks will require time but no cash outlay, and they should be estimated very carefully. They will be valued and added on later.

You can then figure the carcass price by dividing the live weight cost per pound by the dressing percentage.

If the above budget were for a lamb with a live weight price totaling $0.93 per pound and the lamb dressed 50%, the example would yield $0.93 divided by the 50% dressing percentage, or $1.86

per pound of carcass. This example doesn't reflect the cost per pound of edible meat or the expense required to get it in that form.

This analysis is the most powerful tool a farmer can wield. Farmers know what they must charge for a product and must be able to back it up if queried by a customer.

Careful study must go into this budget every time animals are priced because some costs can vary and even be prohibitive. Consider the expense of trucking an animal to a slaughter facility. Let's take the steer from the earlier trucking example: one steer on a pickup truck expensed at $0.50/mile to haul for 50 miles and return. What effect does trucking have on dressed price?

Fifty miles multiplied by 2 (to reflect trucking both to and from the facility), times a rate of $0.50 per mile equals $50.00. Spreading the transportation cost over 1,100 pounds yields $50.00 divided by 1,100 pounds, or 4.5 cents per pound live weight. Considering a dressing percentage of 62%, the carcass price would be affected accordingly. The 4.5 cents per pound live weight when divided by the dressing percentage of 62% equals 7.25 cents per pound of dressed beef.

If two lambs or kids are put on the same truck for the same distance, the math comes out differently. If the lambs were sold for a lamb roast and they weighed 75 pounds each and the predicted dressing percentage is 50%, the cost of trucking on a per-pound basis would be much higher. The trucking expense is the same: 50 miles times two both to and from the facility times the same 50 cents per miles equals $50.00. 2 lambs at 75 pounds each = 150 pounds. $50.00 divided by 150 pounds equals $0.33 per pound of live lamb or kid. If this trucking expense of $0.33 per pound were divided by the expected dressing percentage of a fat lamb, the impact of this trucking would be very great.

Thirty-three cents per pound divided by 50% dressing percentage equals $0.66 per pound of carcass just in trucking, and it does not include the price of the lamb. More lambs on the truck will, of course, lower this expense.

Trucking affects American farmers greatly, and these examples show that it is no different with direct-marketed lamb, goat, or beef.

Consumer Budget

Price analysis leads to a discussion of the consumer's budget. Regarding the slaughter expenses, the best advice is to pay careful attention to them and understand the impact they can have on the price of a small carcass. Some slaughterhouses may establish a single per-head fee for small ruminants, while others may charge a kill fee and then a price for cutting, wrapping, labeling, and quick freezing. Slaughterhouses play a very important role in the image of the marketer. Sloppy writing on labels and bloody packages may negatively impact your customer's satisfaction.

The buyer usually pays for the processing. Often the farmer stands for the killing. The costs of processing and killing demand very close scrutiny, because they may fluctuate considerably, with a difference of as much as $10 to $20 per head just to kill a small animal. That $10 increase on a 35-pound carcass will increase the cost nearly $0.29 per pound. Some farmers have been successful negotiating down the killing fee with their processors by consolidating orders and using the argument that they are steady customers.

Careful attention should also be paid to the skills of the processor, as many butchers lack the experience and knowledge to cut lambs and kids to suit the farmer's customers. Each individual cut of lamb that the customer wants and that the butcher is expected to cut must be agreed upon in advance. Again, the ability to coordinate and manage will be very useful.

Consumers, of course, have a completely different view of the pricing structure. One of their concerns is what the product will cost in a form ready to eat or cut, wrapped, and frozen. The cost of a small animal intended to be served as a meal in its entirety is easily calculated. However, those who buy lambs, kids, and beef for their freezers analyze and calculate cost from a different perspective. This gets complicated when a first-time freezer-lamb buyer asks, "Where's the rest of my lamb?"

Helping the consumer understand the concept of carcass yield will be helpful. Table 7-1 depicts the average carcass composition of various species of meat animals. Table 7-2 indicates the typical yield for lamb.

The point here is that the consumer's price will be affected by every decision the marketer makes.

Table 7-1: Average Compositional Variations among Species of Meat Animals

	Beef	Veal	Pork	Venison	Lamb	Tom Turkey	Broiler Chicken	Farm-Raised Catfish
Live weight, kilograms	550	160	110	70	50	15	2	7
Proportion of Live Weight								
Noncarcass, %	38	46	27	42	48	18	23	37
Carcass skin, %	*	-	5	-	-	9	9	-
Carcass fat, %	17	7	23	10	17	6	7	-
Carcass bone, %	10	15	9	8	10	17	22	12
Carcass muscle, %	35	32	36	40	25	50	39	51
Total	**100**	**100**	**100**	**100**	**100**	**100**	**100**	**100**
Dressing %	62	54	73	58	52	82	77	63
Carcass muscle/bone	3.5	2.1	4.0	5.0	2.5	2.9	1.8	43

*Note: 1 kilogram = 2.2 pounds   *Included with noncarcass component*

With kind permission of Springer Science and Business Media. Muscle Foods: Meat, Poultry and Seafood. *D. M. Kinsman, A. W. Kotula, and B. C. Breidenstein. 1994. Kluwer Academic Publishers, Dordrecht, The Netherlands.*

Table 7-2: Yield for Lamb (%)

Boneless shoulder	15	Loin chops	7
Rib chops	7	Breast	5
Boneless chops	18	Shank	3
Sirloin chops	7	Neck	2
		Flank	3

Yield 62% of carcass weight in packaged lamb with 13% as either stew meat or burger meat.

With kind permission of Springer Science and Business Media. Muscle Foods: Meat, Poultry and Seafood. *D. M. Kinsman, A. W. Kotula, and B. C. Breidenstein. 1994. Kluwer Academic Publishers, Dordrecht, The Netherlands.*

CONCLUSION

Setting a price that will keep a customer happy and satisfy the producer's goals and needs can be defined as the essence of direct marketing. The process starts with a planning budget. Before the producer becomes a direct marketer, he or she must establish a cost per unit of production for the product and then establish a price. The marketer must decide whether to include each of the following budget items in the decision: the product at a premium price, advertising, communications required to make the sale, extra time required to select the animal, delivery, slaughtering, processing, billing (collection and the use of money, as well as the risk of not receiving it), and customer satisfaction calls. These costs must be determined so the product can be fairly and profitability priced.

The goal of every direct marketer should be to provide a wholesome, attractive, and healthful product to the customer. It is assumed that every marketer strives to sell products that were produced and processed in a way that complies with local, state, and federal guidelines. Don't do any direct marketing of meat animals until you have obtained clearance from your state meat inspectors and other agents who are charged with the responsibility to keep America's food safe. Contact your state department of agriculture. First ask them about local and federal regulations. Tell them in writing what you plan to do. Ask them to respond in the same way. This issue gets clouded and complicated very quickly if a farmer starts to market outside the regulations.

Direct marketing to the consumer or other end user provides the producer the very rare opportunity to actually price his or her product. This includes the freedom to make mistakes and to fail. A thorough understanding of the prospective customer and knowledge of the cost per unit of production will help the producer establish a competitive price for the product. After that, careful attention to every detail of each integrated step from production to the sale and after will increase the producer's chance of attaining his or her direct marketing goals.

CHAPTER 8
Hay Marketing

R. Mark Sulc, Marvin H. Hall, and Lester R. Vough

All livestock producers participate in hay marketing at some time, simply because hay is the most versatile and widely used method of storing forage. Managers of even the most efficient and productive pasture-based livestock operations will likely find themselves involved in hay marketing, either by purchasing supplemental feed when homegrown forage is limited or through the opportunity to sell excess forage produced from their meadows.

In this chapter, we discuss hay marketing as a process, a well-planned course of action involving far more than what transpires at the actual point of sale. We will look at hay marketing from the perspectives of both the hay producer/seller and the hay buyer.

THE BUYER'S PERSPECTIVE

Hay making is an expensive and labor-consuming process. The capital cost of the harvesting equipment continues to rise and must be spread across a large volume of harvested forage in order to be cost-effective. On average, at least 500 tons of dry matter must be harvested each year to reduce costs to the point where ownership and replacement of hay equipment can be profitable (4). Unless such an economy of scale can be maintained and justified, the far better option for graziers is to drastically reduce ownership of harvesting equipment. This can be achieved by not replacing aging equipment and by relying on supplemental hay purchases, contracted hay harvesting, or both. Eliminating ownership of harvesting equipment means that the farming enterprise becomes more specialized. Labor formerly spent on hay harvesting is freed up, allowing more focus on the pasture-livestock operation. Careful planning is required,

because this involves greater reliance on outside sources, either for purchased hay or for contract harvesting services. Detailed discussions of contract forage harvesting are available elsewhere (6). The remainder of this section will focus on purchasing hay for livestock.

Defining Hay Needs

Before beginning the search for hay sources, the hay buyer should carefully define the type, quantity, and quality of hay desired. This results in a more focused search. It is usually not cost-effective to purchase hay that exceeds one's needs, but it certainly doesn't make sense to purchase hay that will not meet one's needs. The nutritional requirements of livestock should be the basis for defining hay needs. The "best" hay meets the animal needs at the desired level of animal performance and can be purchased for a reasonable cost. Length of the feeding period, availability and cost of alternative and supplemental feeds, labor availability, feeding facilities and logistics, and hay storage options should all be considered before purchasing hay. These factors affect the optimal hay package size and quantity to be purchased at any given time. How hay is stored is especially important because it affects preservation of the quality and condition of the purchased hay. The package size will usually affect price; larger hay packages are often more economical than small packages, but smaller packages are more versatile.

Evaluating Hay

The hay buyer should obtain a chemical and physical assessment of the hay before making a purchase. The following section describes these methods in more detail.

Laboratory Analysis

This is the most accurate method to assess nutrient composition of any forage and should be the foundation for all price comparisons and purchases. The buyer should know when the hay was sampled and carefully evaluate the test results. These analyses provide the basis for proper supplementation. Once a relationship is established with a seller, the buyer should request analyses that provide the information needed for his or her specific operation.

Field Inspection

Inspecting hay fields just prior to harvest is an excellent method of evaluating a potential hay purchase. When this is possible, the buyer has the luxury of directly evaluating the species mixture, weed and pest problems, and plant maturity. Unfortunately, field inspections are often not possible or practical.

Visual and Sensory Appraisal of Hay

This is the oldest method of selecting hay. It also is very subjective. Visual and sensory assessment of the hay by an experienced evaluator can help identify hay of inferior or mediocre quality, but such evaluations are not adequate alone to predict nutrient composition. Nevertheless, it still behooves the buyer to obtain a detailed physical description of the hay. Consider the following factors when visually evaluating hay:

> **Maturity** — Stage of maturity at harvest is the single most important determinant of forage quality. As the plant matures, fiber levels increase and digestibility decreases. Indicators of plant maturity include: (i) presence and amount of seed heads of grasses or buds and flowers of legumes, (ii) size of stems, and (iii) leafiness. Forages harvested at early maturity will not have many visible seed heads or flowers. The more extended the seed heads on grasses, the more mature it is. Does the hay have large, thick stems that are hollow and brittle, or are the stems soft and pliable? This can affect

palatability and intake. If grass or legume seed shatters out of the bale, then the hay was harvested at a very advanced maturity stage.

> **Species present** — The ratio of legume to grass species affects forage quality. Legumes are usually much lower in neutral detergent fiber and higher in crude protein and calcium, but they may not be much different in energy content or phosphorus level as compared with grasses harvested at similar maturity. Many buyers have specific ideas about the type of grass/legume combination they are looking for. The ADF/NDF (acid detergent fiber/neutral detergent fiber) ratio can provide an indication of the grass/legume ratio. Pure cool-season grasses will typically have ADF/NDF ratios of 0.60 or less, while pure legumes will have ADF/NDF ratios near 0.75.

> **Leafiness and softness** — Leaves are highly digestible and contain a higher concentration of protein than stems, especially as the crop matures. As forage plants mature, the leaf-to-stem ratio decreases and forage quality decreases. Besides being an indicator of maturity, leafiness can also be an indicator of the harvesting conditions. Leaf shatter results if the forage was too dry at raking or baling. Leaf diseases and some insects can also cause premature leaf loss, especially in legumes. Leaf retention and attachment to stems in the bale are important because shattered leaves will likely be lost in the feeding process.

> **Color** — Most buyers want hay that is green. Color can indeed be an indicator of forage quality, but it can also be one of the most deceiving characteristics when judging the nutritive value of hay. The vitamin A precursor in plants is greater when hay is green. A beige color may indicate sun bleaching or leaching of nutrients by rainfall. A brown or black coloration may indicate excessive heating from hay baled too wet (and possibly less available protein) or excessive weathering losses. But

green overly mature hay cured under perfect conditions may have lower nutritive value than bleached hay that is weed-free and harvested at an earlier stage. In summary, consider hay color with caution, and don't let it be the major determinant of the price you pay.

Odor and dustiness — Does the hay have a clean "crop" odor or a musty or burnt odor? Dustiness is often an indicator of molds resulting from hay baled too wet. The impact of these factors on animal performance varies. Horses are especially sensitive to molds and dust. Moldy sweet clover hay should not be fed to any livestock. It may contain high levels of dicumarol, an anticoagulant, which is produced by the action of molds on coumarin, a natural component of sweet clover.

Foreign matter — The presence of foreign matter, such as weeds, dirt, trash, insects, and other contaminants, can reduce forage quality and may reflect on the managerial ability of the hay producer. If weeds are evident, obtain a positive identification of the species to ensure the safety of the hay and to avoid introduction of unwanted weed species to your farm. Some weeds present no problems when fed, but others can be toxic. Most weeds are of lower quality than legumes.

Hay fed to horses should not contain blister beetles. Blister beetles contain a toxin called cantharidin that severely irritates the equine gastrointestinal and urinary tracts. The use of mower conditioners has made blister beetle contamination a great concern because the crushed beetles are retained in the windrow and are picked up during baling. Hay made without mechanical conditioning will have lower risk of blister beetle contamination. Blister beetles are also more prevalent under arid, droughty conditions and in years following heavy grasshopper infestations. Carefully examine hay intended for horses,

especially if it was produced in a dry climate.

Other factors — Consider tightness and condition of the bales. Was hay stored outside, and if so was it covered and elevated off the ground? Were chemical drying agents or preservatives used during harvest? Preservative-treated hay may be softer and leafier because of being baled at higher moisture content; however, if this higher moisture content is retained, it will have a shorter "shelf life" than dry hay. The chemical activity of preservatives is lost with time, and if the hay has not dried out adequately, mold growth can occur once the preservative stops working.

Discovering Reputable Hay Sources

Finding a reliable and reputable source for purchased hay is very important. Although many information sources are available, the recommendation of a friend or trusted acquaintance is especially valuable. Leads and contacts can also be found through local, regional, state, or national hay and forage associations; ads in newspapers and other farm publications; the Internet; and extension service and governmental agency offices. For larger purchases, reputable hay brokers may be available who can quickly and efficiently find hay of the type and amount needed. Local hay auctions may provide initial contacts. Whatever the source, it is always wise to ask questions and seek out references to validate the reliability of the supplier, especially if large quantities of hay are being purchased.

The Internet is increasingly becoming a source of information on hay marketing. Numerous Web sites are available, and new ones continue to appear that cover various aspects of hay production and marketing. Various state and national hay-listing services exist, including extension and departments of agriculture sites. A careful Internet search can identify most of those sites. Web sites may be useful in providing initial contacts for purchasing hay, and many allow buyers to list their hay needs so potential suppliers can contact them. As with any avenue of finding potential hay suppliers, it is important for buyers to make every effort to validate the reliability of unknown suppliers.

Purchasing Locally or From a Distance

Hay of adequate quality to meet livestock needs can usually be purchased either locally or from a considerable distance. Certain regions or areas have the reputation of having higher or lower quality hay than other regions. For example, hay from the western states is generally considered to be of high quality because hay fields are often irrigated and the low humidity and low rainfall often result in good curing conditions. In contrast, haymaking is often a challenge in the humid eastern region of the United States. Biases exist against hay produced in certain parts of the humid region. But regardless of where the hay was produced, how it was produced is a primary factor influencing hay quality. Low-quality hay is produced in areas noted for production of high-quality hay, and very high-quality hay is produced in areas generally noted for low-quality hay. When purchasing hay, the buyer should insist on laboratory test results as well as a physical description of the hay, regardless of where or by whom it was produced.

Locally purchased hay may be less expensive than hay purchased from a distance because of lower transportation costs, but this isn't always the case. Some may feel that local purchasing ensures more control over the delivered product, but this too is not necessarily the case. Whatever the source of hay, it is important for the buyer to ask the right questions, find a reliable and trustworthy source, and clarify the terms of the sale up-front to avoid unpleasant surprises.

Price Considerations

Determining the economic value of hay is extremely difficult. Supply and demand, quality of the hay, reputation of the seller, hay package size,

delivery options, and many other factors affect the price of hay.

Developing a long-term relationship with a seller is one of the best ways to ensure a fair market price. Many sellers are loyal to repeat customers, especially those who purchase large volumes and pay on time. Those relationships can moderate the seasonal and yearly price fluctuations, with the seller not overcharging in lean hay years and the buyer remaining loyal and agreeing to a fair price in years when hay is plentiful and could easily be purchased elsewhere. Developing this type of relationship can provide considerable peace of mind because you no longer have to constantly shop around and deal with unproven suppliers.

The buyer must consider the hay in terms of purchased nutrients to meet animal requirements. Hay should be purchased by weight rather than by the bale. All price comparisons should be on a dry-matter basis, and hays should be evaluated in terms of the value of nutrients per dry pound or dry ton. Thus, the dry matter or moisture content of the hay should be known.

Most people put too much emphasis on the crude protein content of hay. While protein content is important, forages are rarely the cheapest source of protein. The primary reason for feeding forage is to provide the fiber requirements of ruminant livestock. The forage must be sufficiently digestible so animals can eat enough to meet their fiber requirements without depressing overall intake. The higher the nutrient requirements of the animal, the more important forage digestibility becomes. Once the minimum digestible fiber requirement is met by the forage, other sources of nutrients should be considered on a least-cost basis to supplement the nutrients provided by the total forage in the diet.

When comparing hay lots of different quality, hay of higher quality should be worth more. But the decision to purchase higher priced hay of higher quality should be based on the anticipated animal response and the cost of alternative sources of nutrients. For example, lactating dairy animals have a high threshold to increases in forage quality, and thus one can justify a higher price for high-quality hay fed to high-producing dairy cows. In contrast, one can expect less return from feeding high-quality hay to a beef cow having lower nutritional requirements. A software package was recently developed that values feeds for dairy cows based on the anticipated animal response and the long-term average market price of nutrients (5).

THE HAY PRODUCER'S PERSPECTIVE

Producing and selling hay in the northeastern United States is a challenging activity because of unpredictable rainfall patterns. Despite these challenges, many producers in the Northeast continue to find cash hay a rewarding and profitable enterprise. Areas within the region are often deficient in hay, particularly high-quality hay. This rather sizeable hay market provides opportunities to those willing to invest the effort in producing and marketing a quality product beyond their immediate local area. Nevertheless, those contemplating the cash hay business should carefully assess the production risks in their specific environment and develop plans with built-in contingencies to absorb the production problems associated with the regional weather patterns.

Selling Versus Marketing

The terms selling and marketing are often used interchangeably, but they are quite different. Selling can be defined simply as the exchange or delivery of goods and services to a customer for a negotiated price. An example of selling hay is a producer asking a dealer to look at the hay and quote a price: The hay is of average quality, no forage test is available,

and no attempt was made to segregate it according to cutting or quality going into storage. Hauling hay to an auction is another example of selling. In these examples, the selling procedure is relatively easy and simple, but the producer is basically at the mercy of the buyer. Unless the producer is willing to retain ownership of the hay in hopes that the market price goes up, he is faced with someone else determining the selling price.

Marketing is a more involved process than selling, but the rewards are higher prices and repeat buyers who provide a consistent and stable outlet for the hay. With a good marketing strategy, the producer is readily able to sell to the consumer even when supply exceeds demand. An example of marketing hay is when the producer grows several kinds of hay to fit various markets; manages the production and harvesting for high quality; segregates the hay as it goes into storage according to kind, cutting, and quality; and tests each lot for forage quality. This producer establishes contacts with large dairy operations and horse stables and routinely supplies them with top-notch hay. Widespread recognition leads to new customers. Profitability is achieved and maintained as a result of good management and a strong commitment to supplying a product that fits the customers' needs.

After considering the essential elements in developing and maintaining hay markets, each individual should decide whether selling or marketing is the more appropriate strategy in his or her situation. If the primary operation is livestock production, with varying quantities of surplus hay from year to year (for example, hay produced from excess pasture growth), then careful inventory and storage management along with judicious selling is probably the most appropriate strategy to provide additional income. An elaborate marketing effort would not be cost- or time-efficient. Selling hay also may be appropriate for a livestock producer when the hay produced is of higher quality than needed for the livestock on the farm. It may be more profitable to sell the high-quality hay and replace it with hay of adequate quality purchased at lower cost.

Having a cash hay enterprise and marketing hay require a significant commitment of time and resources. Economy of scale is very important, not only for achieving time and labor efficiency, but also because of the high capital cost of equipment ownership. Having appropriately sized equipment is crucial to achieving a profitable cash hay business. Most cash hay producers find it advantageous to mechanize operations as much as possible, although exceptions to this rule exist. Although this chapter does not deal specifically with the labor and equipment needs of a cash hay business, they should be carefully considered in developing a business plan for a cash hay enterprise.

Identifying a Market

A crucial step toward developing a profitable cash hay business is to carefully identify and develop a market. Many people getting into the hay business delay this step until after the hay is produced, but the market should be identified before the crop is even planted. Conduct a market analysis if you are not familiar with potential opportunities. Talk to other growers, extension and agribusiness personnel, and members of organizations such as The National Hay Association (www.nationalhay.org) (2). Market opportunities may be local, regional, or even international. Contact potential customers and ask them what they are looking for in the product and service. Are you in a position to meet their needs? What is the future of the market? Current localized market demands may change abruptly, so try to identify the most stable markets with the greatest potential for growth. Do you plan to sell to a broad clientele, or will you focus on one specific segment (such as dairy or horses)? What types and quality of hay should be produced for those markets? If you planted alfalfa and the market demand is for an alfalfa/grass mixture, marketing may be quite difficult. Are you going to sell by the bale or by the ton? Out of the field or out of storage?

Hay brokers can provide a valuable service because they have the network to move hay quickly and efficiently. Before making an agreement with a broker, seek out sufficient information to ascertain that the broker is reliable and honest. Talk with other hay producers and buyers who use the broker. Make sure you also understand all the specific requirements and details for dealing with the broker, taking precautions to protect yourself from unpleasant surprises.

Hay marketing associations provide opportunities to network and acquire new ideas. These organizations are more prevalent in the western states. If one is not available in your region, consider joining The National Hay Association (2). Although regional or state associations may provide assistance with marketing, it would be a mistake to assume that someone else is going to do the marketing for you. Each individual must take responsibility for marketing his or her own hay in order to have a successful enterprise.

Producing the Product

Anyone can sell hay in years when hay is in short supply, but difficulties arise during surplus years when buyers can be selective. Producing a high-quality product that matches your customers' needs will go a long way toward assuring consistent movement of hay in the marketplace.

Produce a Quality Product

A high-quality product is very important for maintaining customer loyalty. But top quality means different things to different people. For example, some horse owners have personal standards or definitions that differ from mainstream standards of hay quality. Does your market pay a premium for hay of high nutritional value, and does it justify earlier cutting management? Or is market price primarily based on green color and freedom from musty odor and dust? If little consideration is given to high nutritional value in your market, you may want to use a less intensive cutting system to reduce costs.

Buyers have perceptions that some geographic areas are noted for poor-quality hay. If you are in such an area, you will have to overcome that stigma for the product you produce (assuming you do indeed produce a high-quality product). You will have to make forage test results available to the buyer and perhaps make other concessions to gain a new customer. You also should consider techniques that reduce drying time in the field, such as wide swaths and broad windrows, tedding, chemical conditioning, preservatives, or proven harvest equipment modifications. Forced-air drying (barn drying) may be a profitable alternative in some cases.

Have a Consistent and Reliable Supply

To keep customers coming back, you must provide a consistent quality on a continuous basis. If livestock producers have to adjust feeding programs every time they receive a hay shipment, they will find another supplier. You must be able to supply hay for the time period in which the buyer needs it. If the buyer needs a year-round supply and you can provide hay for only six months, then the buyer will likely look for another source. Have a contingency delivery plan to cover equipment breakdowns or other problems.

If a loyal customer routinely needs a type of hay or straw that you do not or cannot grow, consider buying that product for resale as a service to them. Otherwise, the customer may find another producer or supplier who can provide all types of hay and straw needed. Meeting the customer's need to the extent possible is part of a sound marketing strategy.

Put Up a Firm, Solid Bale

Bale size, weight, shape, and type of tying can all affect marketing and price. Tailor bale packages to

Potential Hay Markets in the U.S. Northeast

The horse market is a growing cash hay market in the Northeast, often providing the best opportunity for selling hay at premium prices. Horse owners and trainers can be very particular about the hay they purchase, so the cash hay producer should clearly understand their requirements. Pleasure horse markets usually require (i) small volume purchases at one time, (ii) small square bales that are high quality, and (iii) hay that is free of molds, dust, and blister beetles.

Dairies are another good market potential in the region. Dairy managers are usually highly knowledgeable about the nutrient requirements of their herds and will pay a premium for high-quality hay. Large volumes are often purchased, and opportunities exist to contract for hay delivery well ahead of when it will be needed. Beef producers require varying qualities of hay and may provide an outlet for lower quality hay.

Other potential markets include sheep producers, zoos and other animal farms, feedlots, feed mills, sale barns, and amusement parks. Specialty markets may have unique standards, such as the low bid that meets the minimum quality standards.

The value of hay exports overseas has increased eightfold since 1985 (1). Forage exports are expected to increase as demand for dairy products grows worldwide. Overseas export of double-compressed hay may be a potential market for some.

suit buyers' needs and convenience. One of the greatest deterrents to distant hay marketing in the eastern United States is size and density of bales. Delivery charges are usually based on mileage rather than weight. You may have good-quality hay, but if the bales are light and misshapen and the ties are loose, the market opportunities are limited. Even pleasure horse owners want firm, solid bales, but they usually want 35- to 40-pound bales rather than heavier bales.

Describing and Advertising the Product

Knowing and describing the product adequately is essential in successful hay marketing. In most cases, the more information you provide about the hay, the more likely you are to attract an interested buyer. Determine what information your customers want. The description should be as objective and specific as possible. This will save you and the potential customer time and helps minimize customer dissatisfaction upon delivery.

Useful forms and guidelines for describing and evaluating hay are available through extension offices (3, 7). A good description of the hay is especially important when placing ads. At a minimum, provide the species mix, price, indicators of quality, delivery charges, package type and size, and any restrictions regarding minimum or maximum purchase volumes. Target your advertising to the specific markets that are looking for what you can offer. Know the lingo used by your customers and describe your product using terms with which they are familiar.

It is highly recommended that proper protocols for hay sampling and testing for nutritive value be performed and results be made available to buyers. Hay buyers are becoming increasingly aware of the value of tested hay. Forage testing is the most objective measure of nutritive value. The U.S. Department of Agriculture's new hay marketing guidelines (page 109) were designed to provide descriptors of hay based on nutritive value rather than subjective methods that are too vague. The guidelines are being used by USDA and state department of agriculture reporters for postings in hay reports, in newsletters, and on the Internet to enhance hay marketing nationally. Nationwide uniformity of price information was the goal in establishing these new quality designations.

Pricing Hay

Establishing a price for hay is difficult. A national market price structure for hay does not exist, so effective marketing is very important in getting a good price. Most cash hay producers rely on a combination of experience, assessing the demand, and knowing what others are asking as guidelines in establishing an asking price for hay. Needless to say, hay prices should take into account all costs associated with production, storage, advertising, and hauling the product; therefore, record keeping is very important. Price the product competitively and realistically. The availability and cost of other feedstuffs may affect the price. Some markets provide a greater premium than others for high-quality hay. Know what the requirements are to achieve those premiums and what forage tests are necessary to document the hay quality.

Establishing a base of satisfied repeat customers is critical to the long-term viability of a cash hay enterprise, and fair pricing is part of that process. Many successful cash hay producers maintain a fairly stable price structure for their valued customers. These producers set realistic prices and do not raise them appreciably in response to short-term hay deficits and high market prices. They do this in hopes that their customers will remain loyal and will continue to accept their established price structure when hay is plentiful and prices are low. You may hear reports of very high prices, but it is better to treat your good customers fairly over the long haul than to gain a few high-priced sales of limited volume to people you may never see again.

Revised National Hay Quality Designations/Definitions

USDA and state department of agriculture market news reporters are using revised guidelines in hay market reporting across the country. Reporters use the test measurement most prominent in their area, along with visual characteristics to determine hay quality.

Alfalfa – Alfalfa Testing Guidelines (for domestic livestock use and not more than 10% grass)

Quality Designations	Relative Feed Value	Acid Detergent Fiber	Neutral Detergent Fiber
Supreme*	Over 185	Under 27	Under 34
Premium	170–185	27–29	34–36
Good	150–170	29–32	36–40
Fair	130–150	32–35	40–44
Utility	Under 130	Over 35	Over 44

Grass Hay Testing Guidelines

Quality designations	Crude Protein	
Premium	Over 13	
Good	9–13	
Fair	5–9	
Low	Under 5	

Quantitative factors are approximate, and many factors can affect feeding value.
Based on 100% dry matter.
End usage may influence hay price or value more than test results.

*QUALITY DESIGNATIONS

- **Supreme:** Very early maturity, prebloom, soft, fine-stemmed, extra leafy. Factors indicative of very high nutritive content. Hay is excellent color and free of damage.

- **Premium:** Early maturity, meaning prebloom in legumes and prehead in grass hays, extra leafy and fine-stemmed, factors indicative of a high nutritive content. Hay is green and free of damage.

- **Good:** Early to average maturity, meaning early to mid-bloom in legumes and early head in grass hays, leafy, fine- to medium-stemmed, free of damage other than slight discoloration.

- **Fair:** Late maturity, meaning mid- to late-bloom in legumes, headed grass hays, moderate or below leaf content, and generally coarse-stemmed. Hay may show light damage.

- **Utility:** Very late maturity such as mature seed pods in legumes or mature heads in grass hays, coarse-stemmed. This category could include hay discounted due to excessive damage and heavy weed content or mold. Defects will be identified in market reports when using this category.

Source: Alfalfa Hay 2002. Washington-Oregon-Idaho Market Summary. USDA, Agricultural Marketing Service.
HTTP://WWW.AMS.USDA.GOV/LSMNPUBS/PDF_MONTHLY/WOIHAY2002.PDF

Storing Hay and Managing Inventory

Proper storage and inventory management are an integral part of marketing. In the humid eastern region of the United States, some sort of protected storage facility may even be necessary for those who routinely sell and deliver hay straight out of the field. Protecting the hay from the elements is crucial to maintaining its value. Unprotected hay can suffer significant losses in dry matter and quality, resulting in loss of profits. Storing hay for later sale (such as until late winter) may lead to higher prices; however, this practice should be weighed against the costs and risks associated with retained ownership. A portion or all of the cost of the storage facility should be charged against the hay. Insurance, or possible loss of uninsured hay, should be considered. There is also the "opportunity cost," the amount of interest that could have been realized from immediate sale of hay and investment of the profit. Another potential loss from storing hay is shrinkage (loss of moisture), unless hay is always sold on a 100% dry-matter basis.

Sorting hay by kind, cutting, and quality cannot be emphasized enough. This is crucial if you are to maintain your ability to deliver uniform lots of hay to the customer. Not only should you sort the hay, you also need to have access to the different lots when they will be needed by customers.

If different qualities or kinds of hay are stacked on top of each other, invariably the customer will want the hay on the bottom of the stack. Keeping good records of the location of each lot and its characteristics (such as forage test results, physical description, harvest conditions, type, species, and so forth) is crucial to maintaining hay identity and will ensure a more efficient shipping and delivery operation. Planning deliveries with customers well in advance of when they are needed can greatly simplify hay storage management, but such advance planning with customers is not always possible. So be prepared and equipped to deliver hay of a certain kind and quality on relatively short notice.

The Dilemma of Low-Quality Hay

Hay producers are always going to have some amount of low-quality hay. Disposing of this hay becomes a real challenge and must be anticipated. *Develop a legitimate outlet for this hay. Don't try to hide low-quality bales in the middle of a load, and don't force the buyer to take every bale in the hay shed. Sort hay by quality and price it accordingly.*

Many western cash hay growers have cattle operations that use the lower quality hay. They sell the best and feed the rest. You may be able to do the same, or find customers who can use the lower quality hay. Some areas have a high demand for mulch hay. Excavating and construction companies, highway departments, ski resorts, gardening centers and suppliers, and mushroom growers are some potential markets for low-quality hay as long as it is weed-free.

MAINTAINING GOOD BUSINESS RELATIONSHIPS

Developing and maintaining a good business relationship is important to both buyers and sellers. It requires good communication and having well-defined expectations and agreements before any transaction takes place. For the hay marketer, the most important rule to keep in mind is that *customers are VIPs.* Place yourself in your customers' shoes and treat them as you would like to be treated. Deliver products that meet the customers' expectations of what you promised to deliver. Back up the product with necessary test results. Follow up on all calls quickly and contact customers on a regular basis. Know the customers' needs and listen to their feedback. Keep repeat customers satisfied. Remember the cardinal rule: *The customer is always right.* When you believe the customer is wrong, first go back to the cardinal rule, then deal with the situation as tactfully as possible.

Remember, it is usually worth some sacrifice to develop and maintain a loyal customer relationship and to avoid having a dissatisfied customer spread negative accounts to current or potential customers. For the buyer, it is important to articulate expectations clearly and specifically and treat the seller fairly and honestly. The buyer should communicate any dissatisfaction in a timely and tactful manner. It is also to the buyer's advantage to maintain a long-term, positive relationship with a reliable and honest hay supplier.

The following are important components of hay transactions that will go a long way toward maintaining good relationships between buyers and sellers:

- Availability of an honest, objective, and complete description of the product, including a thorough visual appraisal, type of hay, package size, and form.

- A forage laboratory analysis of the product, meeting the informational desires of the buyer.

- An established fair price of the hay before delivery, with clearly defined terms. Does the price include delivery? Is the delivery price by the load or by the ton? It is best to base the price on tons of dry forage to avoid disputes over bale weights.

- A clearly understood contracted tonnage of hay to be delivered.

- A defined agreement of how disputes over the product will be resolved, established prior to delivery. For example, establish a contingency plan for a sampling protocol and reanalysis of the hay in the event the buyer believes the appearance of the hay does not match the stated quality. Will the hay be sampled in the presence of the seller or delivery person, how many bales will be sampled, where will samples be sent, and who will pay for the reanalysis? Agree on price adjustments based on quality traits and honor those agreements.

- Established conditions for rejection of a delivery, including the obligations of each party.

- An agreement of who is responsible for the hay during shipment.

- Established place, date, and time of delivery. The buyer should advise the seller of any load restrictions on local roads and bridges and provide clear directions to the delivery site.

- Establish who is responsible for unloading and stacking the hay and the procedures to follow. Ensure that the truck can access the site where hay is stacked, even after inclement weather.

- Establish point of sale and method of payment that ensures compliance with any state laws that protect hay producers, buyers, or both. Buyers should promptly pay for any authorized shipment of hay that matches the agreed-upon criteria.

- References provided upon request, for either the buyer or seller.

CONCLUSION

Successful hay marketing provides benefits to both the hay producer and the buyer. There are profitable opportunities for the hay producer willing to invest the effort and resources to develop a sound production and marketing strategy. Hay marketing requires production of a quality product. Increasing yields and reducing production costs will boost profits, but effective marketing is often where the biggest contribution to profit is made, provided there is a quality product to market. On the other side of the transaction, the buyer needs a quality product that meets his or her needs at a reasonable price. Buying hay should be a carefully calculated and planned activity. Planning hay purchases can be useful in reducing costs and improving efficiency in a livestock operation. In this chapter, we have sought to provide ideas that will help hay buyers and hay marketers achieve mutually beneficial relationships.

References

CHAPTER 1: AN INTRODUCTION TO PASTURE-BASED LIVESTOCK PRODUCTION

1. Brighton, D. *Evaluation of the Vermont Environmental Programs in Communities Project.* Kellogg Foundation, 1994.

2. Cambardella, C.A. and E.T. Elliot. Particulate soil organic matter changes across a grassland cultivation sequence. *Journal Soil Science Society of America* 56 (1992): 777–783.

3. Clark, E.A. and R.P. Poincelot, eds. *The Contribution of Managed Grasslands to Sustainable Agriculture in the Great Lakes Basin.* Binghamton, N.Y.: Haworth Press, 1996.

4. Colborn, T., D. Dumanoski, and J.P. Myers. *Our Stolen Future.* New York: Plume/Penguin, 1996.

5. Condon, A. and J. Ashley. Comparative economics of intensive pasture rotation and conventional management practices on Northeast dairy farms: a case study approach. American Agricultural Economists Association Meeting, 1992.

6. Covey, S.R. *The Seven Habits of Highly Effective People.* New York: Simon and Schuster, 1989.

7. Gable, D. *Where Our Food Comes From: The Hendrix College Food System.* Meadowcreek Project, 1986, videocassette.

8. Garry, V. F., D. Schreinemachers, M.E. Harkins, et al. Pesticide appliers, biocides, and birth defects in rural Minnesota. *Environmental Health Perspectives* 104, no. 4 (1996).

9. Gebhart, D. Some agricultural activities are related to carbon dioxide emissions. *Kerr Center for Sustainable Agriculture Newsletter* 19, no. 8 (1993): 1.

10. Greenland, D.J. 1995. Land use and soil carbon in different agroecological zones. In *Soil Management & Greenhouse Effect.* Edited by R. Lal, J. Kimble, E. Levine, et al. Boca Raton, FL: CRC Lewis Publishers (1995), 9-24.

11. Holt, J.S. and H.M. LeBaron. Significance and distribution of herbicide resistance. *Weed Technology* 4 (1990): 141-149.

12. Jackson-Smith, D., B. Barham, M. Nevius, and R. Klemme. 1996. Grazing in dairyland: The use and performance of management intensive rotational grazing among Wisconsin dairy farms. Technical report. University of Wisconsin Agricultural Technology and Family Farm Institute, no. 5, 1996.

13. Lal, R., J. Kimble, and B.A. Stewart. World soils as a source or sink for radiatively-active gases. In *Soil Management & Greenhouse Effect.* Edited by R. Lal, J. Kimble, E. Levine, et al. Boca Raton, FL: CRC Lewis Publishers (1995), 1-7.

14. LeBaron, H.M. and J. McFarland. Herbicide resistance in weeds and crops. An overview and prognosis. In *Managing Resistance to Agrochemicals. From Fundamental Research to Practical Strategies.* Edited by M.B. Green, H.M. LeBaron, and W.K. Moberg. ACS Symposium Series 421. Washington, D.C.: American Chemical Society, 1990, 336-352.

15. Murphy, B. *Greener Pastures on Your Side of the Fence: Better Farming With Voisin Management-Intensive Grazing.* 4th ed. Arriba Publishing, 1998.

16. Murphy, W.M., J.R. Rice, and D.T. Dugdale. Dairy farm feeding and income effects of using Voisin grazing management of permanent pastures. *American Journal of Alternative Agriculture* 1, no. 4 (1986): 147-152.

17. Murphy, W., J. Silman, L. McCrory, et al. Environmental, economic, and social benefits of feeding livestock on well-managed pasture. In *Environmental Enhancement Through Agriculture.* Edited by W. Lockeretz. Tufts University School of Nutrition Science and Policy, American Farmland Trust, and H.A. Wallace Institute for Alternative Agriculture. November 15-17 Conference Proceedings (1995), 125-135.

18. *A Better Row to Hoe: The Economic, Environmental, and Social Impact of Sustainable Agriculture.* St. Paul, MN: Northwest Area Foundation, 1994.

19. Petrucci, B.T. The potential of dairy grazing to protect agricultural land uses and environmental quality in rural and urban settings. In *Environmental Enhancement Through Agriculture.* Edited by W. Lockeretz. Tufts University School of Nutrition Science and Policy, American Farmland Trust, and H.A. Wallace Institute for Alternative Agriculture. November 15-17 Conference Proceedings (1995), 145-150.

20. Pimentel, D., H. Acquay, M. Biltonen, et al. Environmental and social costs of pesticides. In *The Pesticide Question. Environment, Economics, and Ethics.* Edited by D. Pimentel and H. Lehman. New York: Chapman and Hall, 1992.

21. Powles, S.B. and P.D. Howat. Herbicide-resistant weeds in Australia. *Weed Technology* 4 (1990): 178-185.

22. Rayburn, E.B. Potential ecological and environmental effects of pasture and BGH technology. In *The Dairy Debate.* Edited by W.C. Liebhardt. Davis: University of California Press SAREP, 1993.

23. Semple, A.T. *Grassland Improvement*. London: Leonard Hill Books, 1970.

24. Smith, S.F. Economics of rotational grazing generates interest. Agricultural Update. Ithaca, NY: Cornell Cooperative Extension, 1994.

25. United States Department of Agriculture. Agricultural Statistics. Washington, D.C., 1997.

26. Voisin, A. *Grass Productivity*. Washington, D.C.: Island Press, 1988.

27. Washburn, S.P., R.J. Knook, J.T. Green, Jr., et al. Enhancement of communities with pasture-based dairy production systems. In *Environmental Enhancement Through Agriculture*. Edited by W. Lockeretz. Tufts University School of Nutrition Science and Policy, American Farmland Trust, and H.A. Wallace Institute for Alternative Agriculture. November 15-17 Conference Proceedings (1995), 137-143.

28. Winsten, J., D. Hoke, S. Flack, et al. Financial analysis of Vermont pasture-based dairy farms. Paper presented at the Champlain Valley Grazing Conference, St. Albans, Vermont, April 13, 1996.

CHAPTER 3: RESOURCE INVENTORIES IN FARM PLANNING

1. U.S. Department of Agriculture. *National Planning Procedures Handbook, Amendment 2*. Natural Resources Conservation Service, Washington, D.C., 1998.

2. U.S. Department of Agriculture. *National Range and Pasture Handbook*.

3. U.S. Department of Agriculture. Natural Resources Conservation Service, Washington, D.C., 1997.

CHAPTER 4: ALLOCATION OF FARM RESOURCES

1. Barry, P.J., P.N. Ellinger, J.A. Hopkin, et al. *Financial Management in Agriculture*. 5th ed. Danville, IL: Interstate Publishers, 1995.

2. Boehlje, M.D. and V.R. Eidman. *Farm Management*. New York: John Wiley & Sons, 1984.

3. *Financial Guidelines for Agricultural Producers*. Recommendations of the Farm Financial Standards Council. December 1997. Available online at http://www.ffsc.org/guidelin.htm.

4. Ford, S.A. and W.N. Musser. Evaluating profitability of pasture systems. In *Grazing in the Northeast*. Ithaca, NY: Northeast Regional Agricultural Engineering Service, 1998.

5. Kay, R.D. and W.M. Edwards. *Farm Management,* 4th ed. New York: McGraw-Hill, 1999.

6. Musser, W.N., E.M. Dengler, S.A. Ford, et al. Management Intensive Grazing Unit 4: Economics and Environment. Notebook for In-services. Department of Dairy and Animal Sciences. University Park: Penn State University, 1997.

CHAPTER 6: DAIRY MARKETING

Books

1. Armstrong, G. and P. Kottler. *Marketing: An Introduction*, 5th ed. Prentice Hall, 1999.

2. Bailey, K.W. Marketing and Pricing of Milk in the United States. Iowa State University Press, 1997.

3. Blaney, D.P. *The Changing Landscape of U.S. Milk Production*. Washington, D.C.: Economic Research Service, U.S. Department of Agriculture, Agriculture Information Bulletin No. 761, February 2001.

4. Blaney, D.P. and A.C. Manchester. *Milk Pricing in the United States*. Washington, D.C.: Economic Research Service, U.S. Department of Agriculture, Statistical Bulletin No. SB978, June 2002.

5. Elitzac, H. *Food Cost Review*. Washington, D.C.: Food and Rural Economics Division, Economic Research Service, U.S. Department of Agriculture, Agricultural Economic Report No. 780.

6. Heibing, R.G. and S.W. Cooper. *How to Write a Successful Marketing Plan,* 2nd ed., NTC Business Books, 1997.

Web Sites

7. ATTRA: Appropriate Technology Transfer for Rural Areas: http://www.attra.org/

8. CCH International: http://www.toolkit.cch.com/

9. Congressional legislation: http://thomas.loc.gov/

10. Cornell University Program on Dairy Markets and Policy: http://www.cpdmp.cornell.edu/

11. Pennsylvania State University dairy information: http://dairyoutlook.aers.psu.edu/

12. Small Business Administration: http://www.sba.gov/

13. USDA Agricultural Marketing Service Dairy Programs: http://www.ams.usda.gov/dairy/

14. USDA agricultural statistics and publications (searchable): http://usda.mannlib.cornell.edu/

15. Wisconsin University Center for Dairy Profitability: http://cdp.wisc.edu/

CHAPTER 8: HAY MARKETING

1. Morgan, T.H. *U.S. Forage Export Growth Continues.* Paola, KS: Morgan Consulting Group, 1999.

2. The National Hay Association. 102 Treasure Island Causeway, St. Petersburg, Florida 33706. www.nationalhay.org

3. Rohweder, D. *Guide to Evaluating Hay.* Extension Bulletin 3414. Madison: University of Wisconsin Cooperative Extension, 1998.

4. Schnitkey, G. and D. Miller. Costs of alternative forage harvesting and storage methods. In *Efficient Production and Utilization of Forages.* Interdepartmental Series 941. Columbus: Ohio State University, (1994), 61–70.

5. St-Pierre, N.R. Estimating value of nutrients based on market prices of feedstuffs. In *Proceeding of the Tri-State Dairy Nutrition Conference.* 17–18 April 2000. Columbus: Ohio State University, (2000), 137–147.

6. Undersander, D. Contracting for forages. In *Proceeding of the Tri-State Dairy Nutrition Conference.* 17–18 April 2000. Columbus: Ohio State University, (2000), 167–181.

7. Vough, L. *Evaluating Hay Quality.* Cooperative Extension Fact Sheet 644. College Park: University of Maryland, 2000. Available online at http://www.agnr.umd.edu/MCE/Publications/PDFs/FS644.pdf.

Other Books from NRAES

The books below can be ordered from NRAES (Natural Resource, Agriculture, and Engineering Service). Complete book descriptions are posted on the NRAES Web site. The Web page address for each book is given below its description. Before ordering, contact NRAES for current prices and shipping and handling charges, or for a free catalog.

NRAES, Cooperative Extension
PO Box 4557
Ithaca, New York 14852-4557

Phone: (607) 255-7654
Fax: (607) 254-8770
E-mail: NRAES@CORNELL.EDU
Web site: WWW.NRAES.ORG

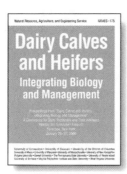

Dairy Calves and Heifers: Integrating Biology and Management

NRAES–175 • 398 pages • 2005
ISBN 0-935817-97-2

Provides practical information on managing nutrition and health to reduce sickness and age at first calving. Includes cost analysis of replacement programs and housing considerations. Intended for dairy producers, their advisors, educators, and agribusiness. Proceedings of a 2005 conference.
WWW.NRAES.ORG/PUBLICATIONS/NRAES175.HTML

Features
- 32 papers
- 70+ tables
- 50+ illustrations

First on the Scene

NRAES–12 • 46 pages • 1989
ISBN 0-935817-06-9

Teaches good decision-making procedures between accident discovery and rescue personnel arrival. Includes decision-and-action diagrams for accidents involving machinery, grain and silage storage facilities, chemicals, and electrocution. Intended for educators, farm families, farm workers, and others.
WWW.NRAES.ORG/PUBLICATIONS/NRAES12.HTML

Features
- Decision trees
- 38 illustrations
- 26,000+ sold

Field Guide to On-Farm Composting

NRAES–114 • 118 pages • 1999
ISBN 0-935817-39-5

Discusses site considerations, raw materials, equipment, recipe making, troubleshooting, mortality composting, and compost utilization. Includes equipment capacity table and material properties. Intended for compost system managers and educators.
WWW.NRAES.ORG/PUBLICATIONS/NRAES114.HTML

Features
- 24 color photos
- 24 illustrations
- 17 tables

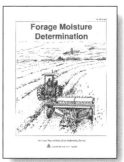

Forage Moisture Determination

NRAES–59 • 28 pages • 1993

The moisture content of stored forage is crucial to successful preservation. This book describes on-farm and laboratory methods of determining moisture content as well as techniques for taking representative samples for testing.
WWW.NRAES.ORG/PUBLICATIONS/NRAES59.HTML

Features
- 11 illustrations
- Equations
- Summary table

more▶

Other Books from NRAES

(continued; see page 115 for ordering information)

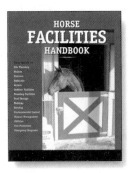

Horse Facilities Handbook

Award Winner
MWPS–60 • 240 pages • 2004
ISBN 0-89373-098-X

Discusses facilities for all phases of breeding, care, and control. Topics include site planning; construction of stables, paddocks, and other indoor and outdoor facilities; environmental control; manure management; bulk feed and bedding storage; fencing; utilities; fire protection; and emergency response planning.

Features
- 111 drawings
- 50+ photos
- Index

WWW.NRAES.ORG/PUBLICATIONS/
MWPS60.HTML

Silage: Field to Feedbunk

NRAES–99 • 464 pages • 1997

Includes practical management information concerning plant and field issues, harvesting high-quality silage, storage, additives management, mycotoxins and spoilage, evaluation, feeding, and silage systems management. Intended for producers, their advisors, educators, agribusiness, animal scientists, and agronomists. Proceedings of a 1997 conference.

Features
- 36 papers
- 50 illustrations
- 117 tables

WWW.NRAES.ORG/PUBLICATIONS/
NRAES99.HTML

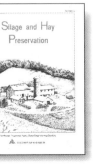

Silage and Hay Preservation

Award Winner
NRAES–5 • 53 pages • 1990
ISBN 0-935817-47-6

This book helps farmers conserve the digestible fiber, protein, and energy of forage and maintain it in a form that animals can use efficiently. Topics covered are the biology of silage preservation, preservation of hay, and additives for silage and hay preservation.

Features
- 15 charts and graphs
- 47 tables
- Equations

WWW.NRAES.ORG/PUBLICATIONS/
NRAES5.HTML

Silage for Dairy Farms: Growing, Harvesting, Storing, and Feeding

NRAES–181 • 450 pages • 2006
ISBN-13: 978-1-933395-06-7
ISBN-10: 1-933395-06-0

Focuses on silage management for improved profitability and reduced environmental impact. Topics include managing nutrients through forages, producing crops for silage, sources and management of mycotoxins, harvesting forages, silage storage strategies, and silage feeding considerations. Intended for dairy producers, their advisors, educators, and agribusiness. Proceedings of a 2006 conference.

Features
- 37 papers
- 86 illustrations
- 126 tables

WWW.NRAES.ORG/PUBLICATIONS/
NRAES181.HTML